NEW FRONTIERS OF CARDIOVASCULAR SCREENING USING UNOBTRUSIVE SENSORS, AI, AND IOT

NEW FRONTIERS OF CARDIOVASCULAR SCREENING USING UNOBTRUSIVE SENSORS, AI, AND IOT

ANIRBAN DUTTA CHOUDHURY

ROHAN BANERJEE

SANJAY KIMBAHUNE

ARPAN PAL

Academic Press is an imprint of Elsevier
125 London Wall, London EC2Y 5AS, United Kingdom
525 B Street, Suite 1650, San Diego, CA 92101, United States
50 Hampshire Street, 5th Floor, Cambridge, MA 02139, United States
The Boulevard, Langford Lane, Kidlington, Oxford OX5 1GB, United Kingdom

ISBN 978-0-12-824499-9

For information on all Academic Press publications
visit our website at https://www.elsevier.com/books-and-journals

Publisher: Mara Conner
Acquisitions Editor: Carrie Bolger
Editorial Project Manager: Judith Clarisse Punzalan
Production Project Manager: Kamesh Ramajogi
Cover Designer: Miles Hitchen

Typeset by STRAIVE, India

Dedication

Dedicated to the Connected Digital Health Research and Innovation Community at TCS Research

Contents

Foreword

> *Deep-learning will transform every single industry. Healthcare and transportation will be transformed by deep-learning. I want to live in an AI-powered society. When anyone goes to see a doctor, I want AI to help that doctor provide higher quality and lower cost medical service. I want every five-year-old to have a personalised tutor.*
>
> ***ANDREW NG***

We are living in the era of a pandemic of noncommunicable diseases, with cardiovascular diseases being the major causative factor to mortality. In the past couple of decades, the advances in cardiovascular medicine have been phenomenal, especially in interventional and preventive cardiology. However, despite all these advances, the burden of cardiovascular disease remains very high. Underdiagnosis, as well as the recurrence of secondary events, is also very high. The spectrum of disease is ever increasing and the population under threat has increased with younger individuals succumbing to lifestyle- and stress-induced factors.

Cardiology is one of the most amenable fields in healthcare to apply artificial intelligence (AI) because of the existence of complex clinical scenarios and multiple comorbidities. This results in an unmet and pressing clinical need for decision support for early recognition of these diseases. AI can tremendously facilitate the continuum of care where early disease detection, future event prediction, and personalized patient care management have the potential to improve the quality of life and clinical outcomes. Current knowledge and practices are based on population-based cohorts. Applying deep learning-based methods to existing and future data collected through modern technologies can open the flood gates for personalization and higher standards of patient care management.

Since this is a pandemic affecting the majority of the population globally, the use of frugal sensors and AI can facilitate screening and management of cardiovascular disease at unprecedented scales, leading to mass personalization.

Let us consider the example of a quite common disease such as hypertension or elevated blood pressure, which unfortunately is the major cause of premature death worldwide. An estimated 1.28 billion adults aged 30–79 years worldwide have hypertension, but about 46% of those adults are unaware that they have the condition. Additionally, only 42% of adults with hypertension are diagnosed and treated. Even after the treatment, approximately only 1 in 5 adults with hypertension have it under control. This is a wonderful use case where unobtrusive sensing can not only help in detection and management, but plays a pivotal role in decreasing the burden of premature deaths. Another compelling use case, I feel, is arrhythmia monitoring and heart failure management.

Some questions looming around are whether AI should replace doctors? Are machines the new doctors? Certainly not. However, doctors who leverage the benefits of technology will certainly be better equipped in making good and appropriate choices as compared to the physicians who do not.

This book is a great step in the right direction and elaborately explains how sensing and AI can help in early disease detection, prediction, and monitoring using unobtrusive sensing. It will help clinicians not only recognize the disease early but also use the data to give personalized and meaningful insights that do not exist today.

Sundeep Khandelwal
Tata Consultancy Services

Preface

Artificial intelligence will not replace doctors. But doctors who use artificial intelligence may replace those who don't

- Berci Meskó, MD, PhD, The Medical Futurist

We, at TCS Research, had started working with connected health applications in 2013. Our first application then was a mobile phone camera-based fingertip photoplethysmogram (PPG) sensing app which can be seen in many fitness tracker apps today. But as we started working in this area, we realized that the real value lies in creation of doctor-acceptable screening applications that can be in the hands of the everyday person. Talking to doctors, we also realized that the biggest requirement for such early screening comes in noncommunicable diseases (NCDs), chronic diseases in general, and cardiovascular diseases. We were surprised to learn that late detection of cardiovascular diseases is the biggest cause of deaths globally, and in many cases, there are very few early signs or symptoms. Hence, we started investigating whether we could sense human physiological signals such as PPGs that carry cardiological markers using easily available sensors (mobile phone and wearable) and then use artificial intelligence (AI) to detect the tell-tale signatures of early disease in those signals. These could be used as an early warning or screening system for seeing a doctor. This book is a result of indepth research in this field by large team of TCS researchers in this area for the past 7 years that synthesizes the knowledge and outcomes of more than 20 papers published in reputed journals and conferences, backed by several filed patents.

In Chapter 1, we introduce the need for cardiovascular sensing. We devote Chapters 2 and 3 to generalized techniques of physiological sensing and AI-based analytics, respectively. In Chapters 4–7, we dive deeper into the specific use of sensing and AI in different disease classes: Arrhythmia, heart blocks, hypertension, cardiac fatigue, sleep apnea, and chronic obstructive pulmonary disease. Finally, in Chapter 8, we take a peek into the future and talk about upcoming sensing, edge computing, and AI techniques that will disrupt and shape the healthcare world and you a glimpse of the life of a patient and doctor, as we foresee, 10 years down the line.

This book would not have been possible without the help of several people. First and foremost, we want to thank Mr. K. Ananth Krishnan, CTO of TCS and Dr. Gautam Shroff, Head of Research, TCS who encouraged us to deep dive into connected health research. We want to thank various scientists and researchers from the Embedded Devices and Intelligent Systems Research Area and Connected Digital Health Research Program in TCS, whose work provided the backbone of the content of this book. We are particularly thankful to Dr. Sundeep Khandelwal, a cardiologist and consultant to TCS, for advising us with his clinical knowledge and agreeing to write the foreword of this book, and to Dr. K. M. Mandana, a renowned cardiac surgeon who made us aware of the real problems faced by doctors in this space. We also thank the editorial staff of Elsevier, whose diligent follow ups and feedback kept us on our toes and helped us in delivering the book on time. Last but not least, we want to thank our families, without whose support and inspiration the writing of this book would not have been possible.

We believe this book will be useful to a wide audience including practitioners and people who are in generally interested in exploring a pragmatic approach to use of sensing and AI to screen for cardiovascular diseases. We want to emphasize that this is not a technology cookbook that is prescriptive in nature, but one that discusses practical considerations for building real systems.

About the authors

Anirban Dutta Choudhury
Senior Scientist, TCS Research, Tata Consultancy Services, India
Anirban Dutta Choudhury received his BE in Electronics and Telecommunication Engineering in 2005 from Jadavpur University, India, and a MSc in Embedded System Design in 2008 from University of Lugano, Switzerland. He has more than 15 years of experience, all in R&D departments in various geographies and scale of companies such as NXP Semiconductors Netherlands, PIXY AG Switzerland, and Tata Consultancy Services India. His research interests encompass noninvasive physiological sensing in digital health and AI-driven sensor signal informatics.

Anirban has more than 40 publications in reputed Journals and Conferences. He has filed for more than 25 unique patents and has 10 patents granted to him.

Rohan Banerjee
Scientist, TCS Research, Tata Consultancy Services, India
Rohan Banerjee has more than 10 years of experience in digital signal processing and machine learning. Currently, he works as a scientist at TCS Research. He received his MTech from the department of Electronics and Electrical Communication Engineering, Indian Institute of Technology, Kharagpur in 2011. His main research area includes signal processing with a strong focus on biomedical signal processing, pattern recognition, machine learning in digital healthcare, and cardiology. Rohan has more than 30 papers published in peer-reviewed conferences and journals and has 5 granted patents.

Sanjay Kimbahune
Senior Scientist, TCS Research, Tata Consultancy Services, India

Sanjay Kimbahune received his BE in Electronics from Amravati University with distinction. He is working as a senior scientist at TCS Research and Innovation. He has experience of 32 years in diverse fields of CTI, video conference solutions and mobile applications for masses. He is passionate about solving problems faced by the masses by developing frugal and scalable technology solutions. He has published 33 papers and has 48 granted patents. His current research focus is on engineering innovative devices for screening cardiovascular diseases in a scalable and frugal way. He is mentoring many startup groups working in the healthcare domain. He was instrumental in developing the mKRISHI platform for farmers. mKRISHI has reached about 1 million farmers as well has received many prestigious awards.

Arpan Pal
Distinguished Chief Scientist, TCS Research, Tata Consultancy Services, India

Arpan Pal has more than 29 years of experience in the areas of intelligent sensing, signal processing and AI, edge computing and affective computing. Currently, as Distinguished Chief Scientist and Research Area Head, Embedded Devices and Intelligent Systems, TCS Research, he is working in the areas of connected health, smart manufacturing and remote sensing.

He is on the editorial board of notable journals such as ACM Transactions on Embedded Systems, Springer Nature's Computer Science and is on the technical program committee of notable conferences such as ICASSP and EUSIPCO. He has filed more than 180 patents (out of which 95+ granted in different geographies) and has published more than 140 papers and book chapters in reputed conferences and journals. He has also authored two books, one on the Internet of things and another on digital twins in manufacturing. He is on the governing/review/advisory board of some of the Indian Government organizations such as Council of Scientific & Industrial Research and MeitY, educational institutions such as the Indian Institutes of Technology, and technology incubation centers such as Technology Innovation Hub (TIH).

Prior to joining Tata Consultancy Services (TCS), Arpan had worked for DRDO, India as Scientist for Missile Seeker Systems and in Rebeca Technologies (previously Macmet Interactive Technologies) as their Head of Real-time Systems. He has a BTech and MTech from IIT, Kharagpur, India and a PhD from Aalborg University, Denmark.

SECTION 1

Sensors, AI and IoT in cardiovascular diseases

CHAPTER 1

Cardiovascular conditions: The silent killer

AI will not replace physicians. However, physicians who use AI will replace those who do not.

- Berci Meskó

It's time to move from reactive sick-care to proactive healthcare by default.

- Koen Kas

Using digital tools to reach patients is no longer a question of "If" it's a question of "How" and it's a matter of "Now".

- Chris Boyer

The coming era of artificial intelligence will not be an era of war, but the era of deep compassion, nonviolence, and love.

- Amit Ray

The promise of artificial intelligence in medicine is to provide composite, panoramic views of individual's medical data; to improve decision making; to avoid errors such as misdiagnosis and unnecessary procedures; to help in ordering and interpretation of appropriate tests; and to recommend treatment.

- Dr. Eric J. Topol

1. Introduction

Since time immemorial, the objective of humankind has been to live a healthy, happy, and prosperous life. Breastfed milk is a complete nutritional package for babies and is the first preventative medicine for infection. Since the beginning of the human race, humans have tried to remain healthy by managing diseases in multiple ways. For maintaining good health, it is imperative that all critical organs are in a healthy state. The critical organs in the human body are the brain, heart, lungs, liver, and kidneys. The heart is the most important organ in the human body, because it is responsible for providing the mechanism to transport glucose and oxygen to these vital organs. Maintaining good cardiovascular health is considered a cornerstone for good health.

New Frontiers of Cardiovascular Screening using Unobtrusive Sensors, AI, and IoT
https://doi.org/10.1016/B978-0-12-824499-9.00001-5

Cardiovascular issues occupy a significantly higher proportion of total mortality from noncommunicable diseases. If these issues are detected early, the chance of recovery and survival increases significantly. Cardiovascular issues are a worldwide problem and hence need "out-of-the-box" thinking. Frugal digital technologies that can be used by semiskilled or even unskilled health workers can be of immense value as scalable and sustainable screening mechanisms.

This book provides insights into the real-world problems in cardiovascular disease screening that can be addressed via the Internet of things (IoT), wearable sensing, signal conditioning, processing, and machine learning (ML). It also provides a practitioner's insight into real-world implementation of such systems.

2. Industrial revolution and Healthcare 4.0

The roots of modern industrial evolution can be traced back in history to the industrial revolution. The first revolution happened in the 18th century in Great Britain when steam-powered engines were used for mass production and transportation. It involved a transformation of manual manufacturing processes to new, machine-assisted manufacturing processes. This was possible because of mechanization, the development of tools, machineries, and the use of steam power and water power. It also resulted in an increase in per capita income in the capitalist society. The roots of development of efficient industrialization and sociocultural changes were seen during this revolution. The enabling factor of the second industrial revolution in the late 19th century was electricity-based mass production systems and assembly lines. Prior to this time, machinery was mostly electro-mechanical. The third industrial revolution was significant as it leveraged the use of electronics and information technology (IT) for further automation of the industry. This happened early in the 21st century when desktop computers and electronic chipsets became commonly available. This was the inflection point wherein early-stage concepts of IoT and ML were introduced. The current industrial revolution is the fourth (I4.0) which is powered by the cyber-physical system. It is characterized with the use of technologies such as artificial intelligence (AI), big data, robotics, the Internet of medical things, and tools such as block chain.

Business is also changing rapidly with this evolution (Business 4.0) [1]. The current revolution is significant as there is a fundamental shift in the way business is conducted. The key highlight of this time is digitization and digitalization. The emergence of technologies such as cheap Internet,

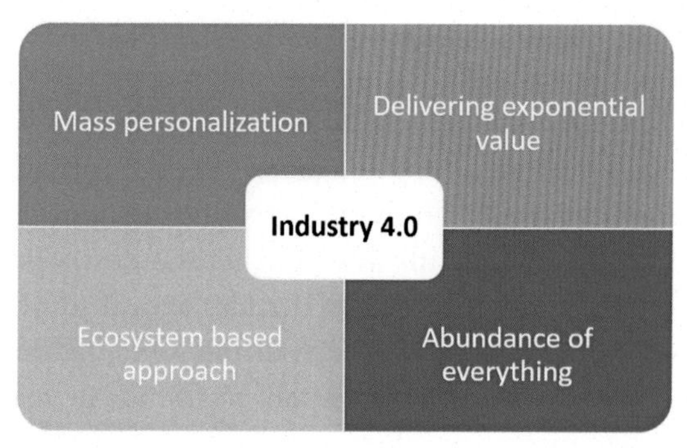

Fig. 1 Basic pillars of Industry 4.0.

cloud storage, and advances in AI/ML algorithms has changed the basic principles or methods of doing business. The emergence of Internet-based companies and new business models has made conventional business models obsolete. Amazon, Uber, and Google are examples of this new way of business. *It is important to understand these aspects of evolution as they are now disrupting the way that health care services are being delivered.*

Fourth-generation industry (Fig. 1) enables best-in-class service using the following pillars:

1. Mass personalization: Learn more and more about an individual and provide the person with what exactly he/she wants, or he/she may want instead. Google searches could be a perfect example of this. The web browser keeps the history of searches, algorithms create a personalization model, and only relevant content is provided to the user.
2. Delivering exponential value: Provide best-in-class services or products at an unimaginably cheap cost. Consumer behavior is rapidly changing. New-age consumers have set their expectations at a very high level. These expectations are often regarding speed, quality, convenience, availability, and experience of service, at extremely competitive prices.
3. Ecosystem-based approach: Previously, organizations used to do everything for a particular product or service. This may include production, warehousing, delivery, servicing, and after-sales support. However, the new approach focuses on collaborating with the best service providers for that task. The outsourcing industry has grown exponentially as a result of this aspect of change. Typically, companies now hold on only to their core business. For example, in the banking industry, the core banking operation is still operated by bank employees. However, ancillary non-core tasks such as payroll, lead generation, customer support, and IT

infrastructure support and management are typically outsourced. This mutually-beneficial relationship enables the prosperity of all the partners in the ecosystem, as each one operates in their own area of specialization, thus as a result, the end consumer receives better service.

4. Abundance of everything: In this hyperconnected world, talent, capital, and capabilities are abundantly available. This industry makes clever use of these enablers. Only a couple of decades ago, if someone wanted to buy a TV set, limited choices were available. One had to visit the store, look at the various models of TVs, talk to friends or read some online reviews and then make a purchasing decision. Now, there are thousands of models of TV sets available online, multiple choices of vendors, price points, services levels, tools to select what fits the consumer's need, and a plethora of review comments. Thus, the consumer can choose the best product as per his/her needs at the desired price point.

Healthcare 4.0, as defined by Siemens Healthineers [2,3], is the model inspired from Industry 4.0. If one tries to map the progress to Industry 4.0 in healthcare [4], it would be (Fig. 2):

1.0 Can be mapped to earlier healthcare where most of the records or operations were manual, including sensing. Examples are tools such as a stethoscope or mercury thermometer.

2.0 Could be mapped with the emergence of electronic health records, including a certain degree of digitization. This could involve the use of devices such as digital thermometers and blood pressure meters.

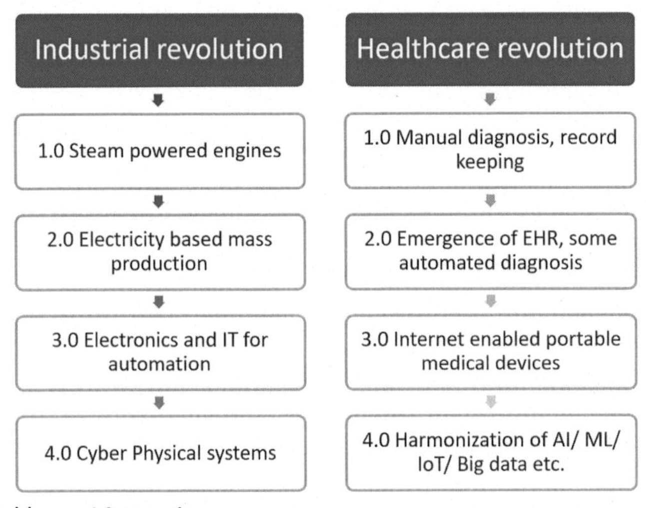

Fig. 2 Healthcare 4.0 mapping.

3.0 Could relate to the emergence of Internet-enabled portable screening or diagnostic devices. These devices could connect to the Internet and upload results or perform remote computations. This could also include the emergence of wearable devices. Examples are loop recorders or pacemakers.

4.0 May be the synchronous harmonization of technologies such as data mining, AI/ML, robotics, IoT, cloud storage, cybersecurity, and the innovative use of supply chains.

The healthcare industry also seems to be embracing the basic tenets of 4.0. The research service, Business Insider Intelligence, predicts that there will be 24 billion IoT devices at the end of this decade. Healthcare 4.0 is primarily about capturing a huge amount of data and putting it to work in an efficient and secure manner. The use of AI/ML, data analytics, and robotics are maturing as technologies. Recently, sensing and information technology is taking center stage and is emerging as a major stakeholder in almost all aspects of health [5]. Medical fraternities are also accepting digital technologies with open arms. Computing is becoming cheaper and more easily available due to cloud storage and broadband Internet. Digital technologies have pervaded in areas such as drug discovery and personalized precision medicine.

One good example of companies embracing the principles of 4.0 is Alive Technologies with their Kardia AliveCor solution [6]. AliveCor is a cheap electrocardiogram (ECG) recording device that can be used by anyone. It can be pasted onto the back of mobile phones, making it accessible all the time. A person can record his/her ECG whenever they want, just by placing their index fingers on the sensing surface. The device can capture single-lead ECG signals and six-lead ECG signals. There are two lead electrodes on the top surface and one electrode at the bottom. The device captures the ECG and sends it to an accompanying mobile phone via inaudible sound waves. The mobile application is embedded with AI-trained models that interpret the ECG on the edge (mobile) itself and can warn users of any potential anomalies. The algorithms learn more about the individual as more and more data is captured. The user can send the data to a physician for appropriate advice.

2.1 PhysioNet challenge

PhysioNet (www.physionet.org) was established under the aegis of the National Institutes of Health (NIH) in 1999. The objective of the institution is to catalyze and conduct biomedical research and education. They offer free access to a variety of physiological and clinical data sets and relevant open source software to conduct analysis of these databases. The platform is managed by the members of Massachusetts Institute of Technology (MIT) Laboratory for Computational Physiology. Community resources are available under three categories:

1. PhysioBank: A collection of anonymized, well-organized, and characterized digital health recordings in time series or other formats. Recorded signals include data related to cardiology, pulmonology, neurology, and other disease conditions. For cardiology specifically, there are multiple annotated data sets related to an apnea ECG database, BIDMC congestive heart failure database, European ST-T database, and long term atrial fibrillation (AF) database. This data is licensed under the open data common attribution license V1.0. These databases may be of interest to the readers who want to undertake hands-on experimentation.
2. Physio Toolkit: A collection of software such as a cardiovascular simulator, cardiac output estimation from atrial blood pressure (BP) waveforms, CVSim—lumped parameter model for human cardiovascular system, ECG kits, and so on.
3. Tutorials: A collection of reference manuals, study materials, and workshop materials.

PhysioNet hosted a challenge in 2017 to develop innovative algorithms for AF classification. A series of 8528 single-lead ECG records, of length between 30 and 60 s, were made available to the research community. These were divided into training data and sample data. The challenge was to classify them into four buckets: Normal sinus rhythm (5154), atrial fibrillation (771), an alternative rhythm (2257), or noisy samples (46). There was participation from many premier academic institutes as well as industries. Researchers were able to achieve overall accuracy of 83% in correctly classifying the ECG signals. A significant body of research is available online for people who are interested in trying some of these research outputs.

The PhysioNet challenge of 2022 was renamed the "George B. Moody PhysioNet Challenge" to celebrate the contribution of Dr. George B. Moody in cardiology. The provided data is from a CirCor Digiscope dataset comprising of phonocardiogram (PCG) signals from four main auscultation locations on the body using a digital stethoscope. The database has 5272 recordings and is made available by Universidade Portucalense and Universidade do Porto in Portugal. Each of the heart sounds is annotated by experts by considering its timing, shape, and pitch. The challenge is to develop algorithms to identify cases where a murmur is present or absent, or where signals are not clear.

This book will discuss how some of the aspects of Healthcare 4.0 can be leveraged to have a frugal and scalable cardiovascular screening system with focus on ML.

3. Chronic diseases and cardiovascular issues

Understanding the biological basis of cardiovascular issues is very important. In-depth understanding of these issues can potentially enable researchers to analyze the collected sensor data more effectively. The following is a description of a few critical terminologies, issues, and a brief overview of cardiological conditions.

Heart-related conditions broadly come under the category of chronic diseases. There are various definitions for chronic diseases and conditions. The Commission on Chronic Illness, United States [7] has defined framework to address chronic illnesses at scale. The Center for Disease Control (CDC) defines chronic disease as "conditions that last for more than a year and needs ongoing medical attention or limit daily living or both" [8]. They are permanent, leave residual disability, are caused by nonreversible pathological alterations, require special training of a patient for rehabilitation, and may require a long period of supervision, observation, or care. As a rule of thumb, most of the noncommunicable diseases (NCDs) are chronic. The disease pattern is changing rapidly as a result of changed behavior of people, including approaches toward nutrition, physical activity, and climate change. The health goals of society are also evolving with the advent of technology. With better diagnostic and medication facilities, the average lifespan of society is increasing. However, increased life span is complimented by an exponential rise in NCDs, especially toward the end of life.

In the current age of industrialization, more and more tasks are being automated. In order to do things efficiently, humans are becoming more machine-dependent, which is drastically reducing physical activity. Industrialization has also resulted in an exponential increase in contamination of natural resources. This includes drastic degradation of air quality, water contamination, and the greenhouse effect. The rate of change of degradation is much higher than the previous few decades. The problem is further compounded by the availability of cheap and attractive junk food. With the advent of medical technologies, communicable diseases are more or less addressed. However, the above factors have significantly boosted the mortality from NCDs.

The National Institute of Health Consortium in 2001 defined a biomarker as "a characteristic that is objectively measured and evaluated as an indicator of normal biological process, pathogenic process, or pharmacological response to therapeutic intervention" [9]. In this book, there is a focus on capturing, analyzing, and interpreting cardiology-related biomarkers.

4. Cardiovascular diseases: A silent killer

The World Health Organization (WHO) [10] notes that NCDs are the leading cause of death globally, and one of the major health challenges of the 21st century. NCDs are estimated to account for 71% of 57 million predicted deaths by 2025. Within the overall figure of 71% deaths, approximately 15 million people will be in the age group of 30–68 years. These deaths may be premature and be more prevalent in low- or middle-income countries.

An action plan by WHO lists the first global target of a 25% relative reduction from NCDs. After a careful look at the global mortality as at 2016, NCDs contribute to 57% of total deaths, of which cardiovascular diseases are a major contributor at 31%, neoplasms are 16%, other NCDs are 15%, chronic respiratory diseases 7%, and diabetes at 3%. Thus, it is clearly evident that cardiovascular diseases need urgent attention. WHO has suggested that early detection methods include screening, treatment, and palliative care, as key components of the response to NCDs. These deaths could potentially be prevented if appropriate and timely treatment can be given. Considering the population explosion, resource scarcity, and geographical spread, it is imperative to use digital technologies to tackle this issue on a wider scale. A report by Vaughan et al. [11] indicates that the younger generation, and those living in rural areas, need greater attention in regards to reducing the overall cardiovascular morbidity and mortality.

There are conventional tools and technologies that are currently used for screening cardiovascular issues. This includes devices such as the computed tomography (CT) scan and ultrasounds, and processes such as coronary angiograms. These resources and skills are available at a few select places and thus are out of reach of people at the far reaches of society as their economic conditions do not allow them to take these tests and receive appropriate and timely treatment.

4.1 Factors enhancing cardiac risks

As per WHO [10], risk factors for cardiac problems are classified into two categories:

- Nonmodifiable factors: These could include defects by birth, permanent, nonreversible changes in the functioning of organs. An example could be type 1 diabetes.
- Modifiable behaviors: Tobacco consumption, primary and secondary smoke exposure results in approximately 7.2 million deaths per year [10]. Next, excess salt or sodium intake results in approximately 4.1 million deaths. 3.3 million deaths result from alcohol consumption, and surprisingly, approximately 1.6 million deaths come from insufficient physical activities. All these factors can be addressed by education, behavior change therapy, and encouragement by family members.

In addition to this list, there are metabolic risk factors. These factors are resultant of inappropriate metabolic functions and contribute to raising the risk of NCDs. These include being overweight or obese, consistently raised blood pressure, hyperglycemia (increase in blood glucose levels) and hyperlipidemia (high level of fat in the blood). Any one, or a combination, of these factors poses high risk.

Sustained raised blood pressure is one of the major risk factors of CVDs. In 2015, globally, 1 in 4 men and 1 in 5 women had raised blood pressure. Smoking is strongly linked to heart disease and stroke. Smoking causes thickening and narrowing of blood vessels, increases the build up of plaque, and makes blood thicker. Diabetes, obesity, and air pollution are other factors as well.

WHO nominates that cardiovascular diseases can be classified as (Fig. 3):

- Coronary artery disease: Coronary arteries are blood vessels supplying blood to the muscles of the heart. This class refers to the disease occurring because of blockage of these arteries. This disease will be covered in detail in Chapter 5.
- Cerebrovascular issues: The human brain is approximately 2% of the body weight. It consumes approximately 20% of the total energy [12]. This class of disease corresponds to issues related to blood-supply vessels to the brain.
- Peripheral artery diseases: This class of disease is related to blood vessels supplying blood to the arms and legs.
- Rheumatic heart disease: This disease is a result of inflammation caused by untreated or improperly-treated streptococcal infection. Inflammation can

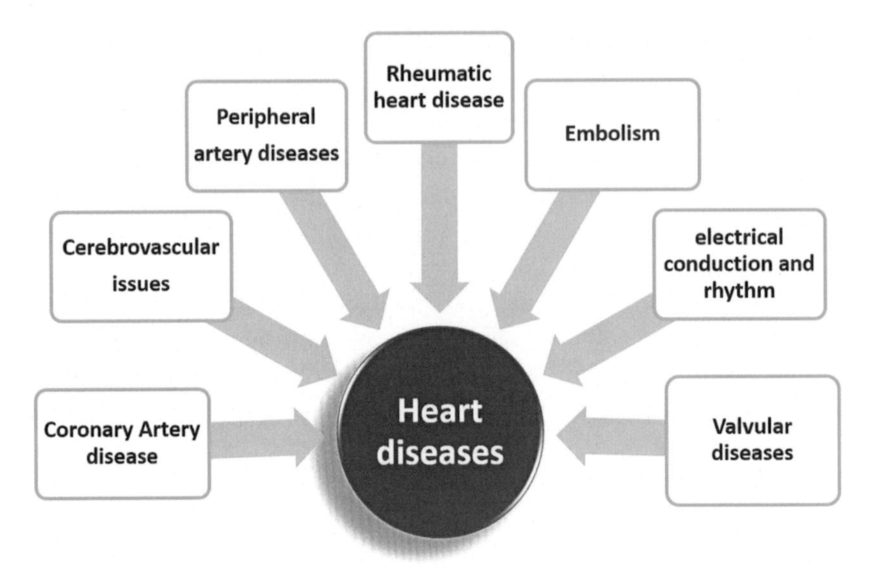

Fig. 3 Categorization of heart related issues.

cause damage to heart tissues, especially the valves. These valves lose elasticity, making it difficult for the normal operation of the heart. Rheumatic fever typically occurs in children aged between 5 and 15 years. Its effect is felt after a long time.

- Congenital heart disease (CHD): The term "congenital" means defects that are present from birth. This refers to the set of heart-related diseases that affect the normal functions of the heart. CHD affects approximately 0.8%–1.2% of live child births worldwide [13]. With advances in neonatal cardiac screening technologies, deaths attributable to CHD are decreasing, however much work needs to be done in developing and underdeveloped countries. CHD comprises of defects that affect or interfere with venous drainage, septation of cardiac elements, their sequence, and functioning of the heart [13].
- Deep vein thrombosis (DVT) and pulmonary embolis DVT refers to the formation of a blood clot in the deep vein. It usually occurs in the legs. DVT is a preventable and treatable disease. It includes venous thromboembolism and pulmonary embolism. It is estimated that 1 per 1000 people experience some sort of embolism in their lifetime [14]. Clots are formed if the balance between coagulation mechanism and anticoagulation mechanism is disturbed. The Virchow Triad describes risk factors for DVT. These are venous stasis, vascular injury, and hypercoagulability.

When the clot travels and lodges in the lungs, it is termed a pulmonary embolism. DVT can be dangerous if the clot travels to the lungs or heart. Aortic stenosis is the most common issue apart from hypertension and coronary artery disease [15].

Some of the other important heart disorders are:

- Abnormalities in electrical conduction and rhyth The cardiac conduction system (CCS) of the heart plays a very important role in maintaining its function. The heart produces approximately 2.3 billion heartbeats during a lifetime [15]. The details of the heart's electrical system are discussed in Chapter 2. The pathway comprises of subsystems such as the sinoatrial node, atrioventricular node, bundle branches, and Purkinje fibers. These units can be viewed as an electrical circuit. Any defects in this system result in conduction abnormalities and rhythm fluctuations. An electrophysiologist can analyze the electrical signals at predefined locations and take corrective actions. This may include ablation of certain paths in the conduction system.
- Malfunctioning of parts such as valves: Due to congenital defects or because of factors such as infection, certain parts of the heart might not work as desired. Valvular heart disease (VHD) accounts for approximately 10%–20% of all cardiac surgeries in the United States [16].

Some of the important cardiac and related diseases are discussed in detail below. This will give a brief overview from a biological perspective. For some of the issues discussed below, subsequent chapters will discuss the IT/sensor and AI-related perspective for screening of these diseases.

4.2 Heart rate, arrhythmia, and atrial fibrillation

Heart rate is also known as the pulse rate. It is the number of times the heart beats in 1 min. The normal range for adults is 60–100 beats per minute. This rate may vary depending on multiple factors such as physical exercise or stress. When the heart rate increases, the rate of blood circulation increases, and thus the body receives more oxygen and nutrients. The beating of a heart follows a particular pace and rhythm. If the heart beats faster than normal, it is referred to as tachycardia, and if it beats slower, it is referred to as bradycardia. A normal heart follows a rhythm, namely cardiac rhythm or sinus rhythm. Any nonconformance to sinus rhythm is referred to as arrhythmia [17]. Symptoms of arrhythmia include accelerating or deaccelerating heartbeat speed. Sometimes beats are skipped. It might result in a feeling of light headedness, chest pain, or sweating.

Atria are the upper chambers of the heart. Sinoatrial (SA) node impulses initiate the contraction of atrial chambers. If cardiac events related to polarization and depolarization of atriums go out of rhythm (time sync or level of exciting signals), it results in an irregular, often rapid heart rate. This asynchronicity could result in incomplete filling or emptying of atriums. These types of disorders are collectively called atrial fibrillation (AF). It also could lead to blood clots that may travel to coronary arteries and result in embolisms. Patients suffering from AF often feel palpitations, a sensation that their heart is racing or flip-flopping, or a reduced ability to exercise [18]. Chapter 4 will describe methods and algorithms to determine abnormal heart rhythms by looking at ECG and photoplethysmogram (PPG) data.

4.3 Coronary artery disease and ischemic heart diseases

The cardiac muscles are one of the strongest muscles in the body. They have a key role of sending the blood (cardiac output) across various parts of the body at an appropriate pressure and volume. The blood is a carrier of oxygen and nutrients on a forward path, and it carries back carbon dioxide and impurities on its return path.

Exercise leads to the following three things:

- The nervous system stimulates the entire circulation system.
- Blood pressure rises.
- Cardiac output increases.

The heart is also trained by regular exercise. For a normal person, when strenuous exercise is performed, the typical cardiac output is approximately 4–5 times the normal cardiac output. For trained athletes, it is much higher. There is a complex process of regulating the oxygen levels by increasing cardiac output using control mechanisms such as vasodilators, sympathetic vasoconstrictor nerves, adenosine triphosphate (ATP) levels, potassium ion levels, lactic acid levels, and so on.

The heart muscles play a key role in ensuring the optimum delivery of blood across body parts. To perform this task, the muscles also need an oxygen supply. Although the heart delivers blood to the body, its muscles are not conditioned to absorb sufficient oxygen from the blood within the heart. Heart muscles receive their blood supply through the coronary blood supply system present in its periphery. It comprises of coronary arteries and veins strategically placed on the heart muscles. For the heart muscle to function correctly, the coronary circulation system needs to be adequately supplied with blood. However, there are a few conditions in which the blood supply to coronary arteries is reduced. One such condition is atherosclerosis. For people with high body mass index (BMI) or chronic hypertension or due to some genetic defects, a large quantity of cholesterol is deposited beneath the endothelium of coronary arteries. Additionally, calcium is deposited, resulting in atherosclerotic plaque. This plaque impedes the blood flow through these arteries. Atherosclerotic plaque can cause a local blood clot, termed a thrombus. These blocks are typically in the beginning portion of the coronary arteries. Acute coronary occlusion can happen for patients with atherosclerotic blockages when plaque comes into direct contact with flowing blood. Due to an uneven surface, blood platelets stick to the clot and fibrin is deposited. This cycle repeats until it completely occludes the vessel [19]. If bloodflow stops or reduces greatly because of coronary occlusion, the relevant portion(s) of heart muscle do not receive enough blood supply and cannot sustain cardiac muscle function. In other words, it is said to be infarcted. Infarcted tissues can either recover or become dead, forming a scar. It takes months or even years for a body to replace the scarred tissues.

The most common method to detect coronary blockages is by performing a coronary angiogram. This process involves coronary catherization. In this process, a physician inserts a catheter in the femoral or radial artery. A catheter is guided up to the coronary arteries, an X-ray dye is injected, and a series of X-ray images are taken. Looking at the images, the state of the blood flow can be visualized. This also indicates the presence of any blocks. The same catheter can be used for corrective actions such as inserting stents to widen the coronary arteries, thus improving the blood flow to the heart. Chapter 5 will provide insights on alternate noninvasive mechanisms of analyzing ECG, photoplethysmogram (PPG), and PCG data to discover blockages in the heart.

4.4 Cardiac fatigue and hypertension

Let us first understand the concept of muscular fatigue. In a human body, there are three types of muscles:

- Skeletal muscles, mainly muscles connected to bones and responsible for locomotion.
- Smooth muscles, which are muscles such as the intestine.
- Cardiac muscles, the muscles that form the structure of a heart.

Muscles derive energy from ATP (adenosine triphospate). A very high-level process of the supply of energy to muscles is through the interaction of creatine and ATP. ATP is converted into adenosine diphosphate (ADP) and energy is released. An important byproduct of this process is the accumulation of lactic acid in muscle cells. Lactic acid accumulation results in the feeling of fatigue. The rate of removal of lactic acid and energy production decides how much time a muscle can work without getting tired. Muscle training plays an important role in improving longevity. Depending on the rate of energy, aerobic or anaerobic respiration is used. Mitochondria play a pivotal role in the energy generation process.

The relative density of mitochondria denotes the capacity of the muscular activity without getting tired. It is interesting to note that every muscle becomes tired after use, but cardiovascular muscles appear to work continuously. The trick is in the following two factors: The mitochondrial density in cardiac muscles is much higher compared to other muscles; and although it may appear that heart is working continuously, it follows a cardiac cycle where only a particular set of muscles is used in a phase [20]. Thus, the remaining sets of muscles have a micro rest. The typical cardiac cycle lasts for 0.8 s and has four phases. A complete cardiac cycle is explained in Chapter 2.

4.4.1 Cardiac fatigue

Fatigue is a common symptom that could be a result of one or multiple disorders. These disorders could be related to cardiovascular issues, viral fevers, and so on, and are often ignored. However, it could be an early warning signal for cardiovascular issues such as coronary artery disease or valvular disease.

Cardiac fatigue is defined as a state in which the cardiac muscles do not receive sufficient nutrient supply and there is an accumulation of lactic acid in the muscle. It is often ignored because the apparent symptom is feeling tired. There are so many reasons for feeling tired, hence these body signals are often ignored by patients. In many cases it leads to deterioration in the quality of life. From a cardiovascular perspective, fatigue could be a result of loss of efficacy of cardiovascular muscles. The New York Heart Association (NYHA) classifies the subject in four distinct classes based on their symptoms. One of the standard tests to assess cardiac fatigue is cardiopulmonary exercise test (CPET). In this test, a patient's ECG and breath samples are collected while the subject is performing a set of predefined activities. Analysis of this data provides an assessment of cardiac fatigue level.

A smartwatch-based platform can be used for unobtrusive monitoring of cardiac fatigue [11], arrhythmia [21], atrial flutter [22], and palpitations [23]. Derived features from PPG signals such as heart rate, breathing rate, and breathing signal power could be used as biomarkers for cardiac fatigue [24]. Chapter 6 describes cardiac fatigue modeling in detail.

4.4.2 Hypertension

Hypertension is one of the most widespread chronic diseases around the globe. The number of people with hypertension has doubled to 1.28 billion, of which 700 million people remain untreated [25]. Hypertension is defined as abnormally high blood pressure. Blood pressure is measured in terms of systolic and diastolic values. Systolic pressure is a function of cardiac output (left ventricle contraction), whereas diastolic blood pressure is a function of the total peripheral resistance or arteries when aortic valve is closed. According to Joint National Committee 7 (JN7), the range of normal blood pressure is 120 mmHg for systole and 80 mmHg for diastole. Anything above 140 mmHg of systolic and above 90 mmHg of diastolic pressure is termed hypertension. The band in between is termed prehypertensive [26]. Mean atrial pressure is the weighted average of systolic and diastolic pressure that maintains blood flow to the tissues, whereas mean systemic pressure is a pressure that maintains the venous return to the heart. Continuously raised blood pressure often results in several complications. These could be microcapillary damage including rupture and thrombosis. The clotting of blood in critical organs can lead to complications such as cardiac arrest or brain stroke. One needs to be very careful in managing blood pressure. The assessment of vascular function is carried out by looking at parameters such as endothelial function, smooth muscle function, arterial stiffness, and certain biomarkers [27]. These markers are also used to predict the degree of atherosclerosis, efficiency of interventions, and outcomes in CVD. The typical diet in Asian countries has more salt intake than required. This is considered one of the risk factors for people with hypertension. A lot of education is required, especially in the remote parts of the world, where many people do not know that they might be suffering from hypertension for a long time. This could be fatal in the long run.

Chapter 6 describes electrical modeling of the cardiovascular system. This modeling work involves the use of PPG to simulate a digital twin of the heart, with special focus on simulating conditions related to hypertension.

4.4.3 Regulation of blood pressure

The body has a control system that tries to maintain the blood pressure within limits. There are various subsystems/parameters involved in maintaining blood pressure. Blood pressure is controlled by the following three mechanisms:

- Baroreceptor reflux: Baroreceptors are located in the aortic arch and carotid sinus. They act like a strain gauge. As the blood pressure increases, the volume of vessels [sinus] increase. This induces a strain on baroreceptors which is sensed by the brain through nerve fibers [9th and 10th cranial nerve] and brain enables modulation of SA node to adjust the SA node firing frequency.
- Renin angiotensin aldosterone system (RAAS): This is a slower control mechanism but a long-term aspect for blood pressure regulation. In response to changes in blood pressure, kidneys produce prorenin, which is converted to renin and released into the blood. It interacts with angiotensinogen produced by the liver, and is converted into angiotensin I. It is further converted into angiotensin II. Angiotensin II affects actions such as altering vasoconstriction, increasing sodium reabsorption, releasing aldosterone, and also acts on the brain parts to stimulate thirst. This action alters blood pressure. Overactive or malfunctioning RAAS can lead to hypertension.
- Antidiuretic hormone: ADH, also known as vasopressin, is a hormone released by the hypothalamus. It is released under conditions such as an increase in plasma osmolality or reduction in blood volume. ADH acts on nephrons to modulate permeability, thereby increasing intravascular fluid volume, an increase in venous pressure, and thus increasing cardiac output.

Blood pressure is a function of the following three main factors: Cardiac output (the normalized amount of blood ejected by the heart each minute), blood volume (total amount of blood in one's body), and resistance offered by the circulatory system. This includes factors such as the flexibility of arteries, mean diameter, accumulation of fats, and blood viscosity. Various medicines modulate these three parameters to manage blood pressure.

Persistent high blood pressure can damage the walls of the arteries. It can cause a heart attack or kidney disease. It might also result in an aneurysm or bubbling of arteries. If this bubble breaks, blood flows out and may form a clot in critical organs. Thus, it is very important to treat hypertension at an early stage.

5. Sleep apnea and pulmonary conditions

Sleep is an important part of human life. It helps in rejuvenating, recuperating, and cleansing the human body. On an average, humans spend one-third of their life sleeping. A sound sleep is essential for good human health. However, lack of sleep can give rise to multiple complications in the long run. Sleep disturbances due to respiration blockage are often termed obstructive sleep apnea (OSA). The Oxford Learners dictionary defines apnea as "A condition in which someone stops breathing temporarily while they are sleeping."

The two air pathways, one through the nose and the other through mouth, merge in the neck area. Typically, with an increase in weight or age or other factors, this area becomes flabby. When one is awake, the muscles are tight but when one is sleeping, these muscles become loose and sometimes obstruct the air pathway. As lungs create a negative pressure for breathing and the pathway in the neck is blocked, the oxygen level in the blood starts dropping. If it falls below the threshold, the brain sends an arousal signal so the person momentarily wakes up. This process sometimes tightens the muscles and helps in clearing the air pathway. The degree of obstruction is measured in terms of number of these events per hour. Prolonged duration of hypoxia affects the organs critically and a person may collapse in a state of syncope.

OSA was identified as impacting many health outcomes and physiological processes such as blood pressure, cardiovascular issues, and others aspects such as the reduction of neurocognitive functions [28,29]. It is also associated with acute blood pressure elevation, cardiac remodeling, and insulin resistance. The major risk factors are obesity, aging, gender, menopause, and ethnicity.

OSA can be diagnosed by a polysomnogram test. This involves recording of multiple sensors such as ECG, PPG, and measurement of airflow. The degree of apnea is measured in terms of the apnea-hypoapnea index (AHI). It represents an average number of apneas or hypoapneas per hour of sleep. Five AHI events per hour are considered normal, up to 15 are moderate, and up to 30 are considered as severe.

Undetected and long-term OSA can result in conditions such as hypertension, stroke, coronary artery disease, arrhythmias, and congestive heart failure (CHF). In patients diagnosed with CAD, the prevalence of OSA ranges from 30% to 57% [29]. It is also related significantly to CHF and arrhythmias. Treatment of OSA typically involves using positive air pressure (PAP) devices. These devices deliver pressurized air while the patient is sleeping. There are two major types of devices: Constant positive air pressure (CPAP) device that delivers constant air pressure while inhaling and exhaling. This method is a bit uncomfortable for patients, especially during exhaling. Another method, a BiPAP machine, delivers reduced pressure while patient is exhaling. This offers some comfort to patients. Chapter 7 describes prediction models for sleep apnea derived from PPG, ECG, EEG, and some breathing parameters.

6. Need for early screening and diagnosis

Medicines have also evolved over time. Preventive and social medicine (PSM) talks about four stages to achieve overall wellness [30]. The first stage of medicine is curative medicine. This approach is approximately 100 years old. The primary objective is to remove disease from an individual. It involves diagnostic techniques and treatment. The second stage is preventive medicine. It is applied to healthy people. Its primary objective is the prevention of disease and promotion of health. It involves various mechanisms such as vaccinations and mass screenings. The third stage is achieving social wellness, and the proposed last stage is spiritual wellness. This is not related specifically to a religion, but it is a sense of achievement, sense of consciousness, self-awareness, being in harmony with nature, and enjoying life to the fullest.

In normal circumstances, people visit a hospital when they are sick. There could be a pattern of diseases for people visiting a hospital. However, if one views the community as a whole, the pattern of diseases is quite different. A very large proportion of diseases are hidden from physicians or common people; however, they are present in large numbers. It may be that these diseases are initially asymptomatic or symptoms are ignored, being not so painful or troublesome. This phenomenon is often referred to as the "tip of the iceberg" [30]. Detection, treatment, and control of these undiscovered diseases (the hidden part of iceberg) is a big challenge to the medical fraternity. It is a known fact that early detection of diseases significantly increases the chances of recovery or even survival. As time is of the essence, screening plays an important role in bringing the potential diseased population in touch with the medical fraternity. The content of this book describes mechanisms for digital screening of cardiovascular diseases. Hence it is particularly important to understand the overall concept of screening in detail.

Screening is different from diagnostics or periodic health checks. Health checks or diagnostics are initiated by an individual when he/she is sick or having symptoms, whereas screening is an activity initiated by an interested set of people for the wider group. The primary intention is to bring at-risk cases to the surface for people who are apparently normal. Fig. 4 describes

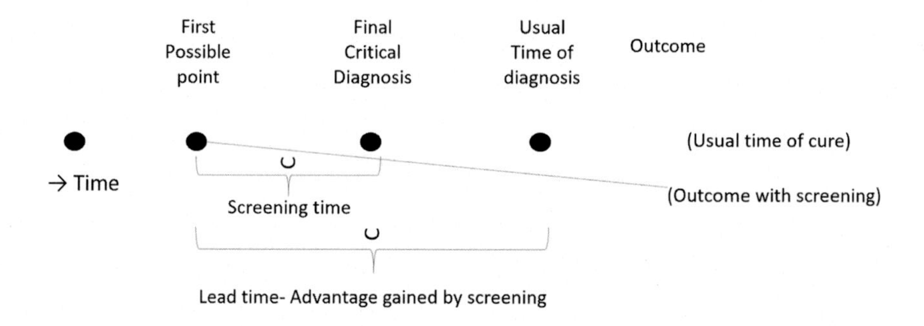

Fig. 4 Model for screening program.

model for screening programs. The following are some of the important aspects of screening [30]:

- It is a moderately accurate method and inexpensive.
- It is used for case detection or prescriptive screening that does not arise from a patient's request.
- It could be undertaken for research purposes. This could pave the way to develop accurate diagnostic mechanisms,
- It could be done en masse or in selective high-risk cases or could be multiphasic, involving multiple tests for a single person.
- It should be acceptable to the end user considering any biases or phobias.
- It should be repeatable with reasonable consistency

The other important aspect of screening is validity of the results. Validity comprises of two components: sensitivity and specificity. Formulas are expressed in terms of a, b, c, d where a is true positive, b is false positive, c is false negative, and d is true negative.

$$\text{Sensitivity} = a/(a + c)\ x\ 100$$

This term was introduced by Yerushalmy in 1040. It is the statistical index of diagnostic accuracy. In other words, it is the ability of the test to correctly identify people with diseases.

$$\text{Specificity} = d/(b + d)\ x\ 100$$

It is the ability to correctly identify nondiseased people.

Predictive accuracy:

It indicates the probability of positively identifying a patient with positive results as the disease in question.

7. How sensing and AI can help

In the past decade, mobile phones have pervaded an exceptionally large portion of the population. People use mobile phones not only for voice communication, but also in data-centric applications such as gaming and health monitoring. In fact, almost half of the world's population is connected through mobile devices. There is another silent revolution happening in the form of IoT, the Internet of Things. A few important sensors are discussed in detail in Chapter 2. The third dimension is the emergence of novel AI techniques. AI brings out insights in the pool of data that are typically not easily discernable to the human mind. Some of the signal processing and AI techniques are described in Chapter 3.

There are few practical problems/challenges with respect to AI and ML [31]. They are:

- Trust and reliability of AI-based inferencing: All algorithms have a certain sensitivity and specificity. They are not 100% accurate. The model heavily depends on the quality of the data on which they are built. One needs to be very careful while using these inferences, especially when they are being used to diagnose or treat people. The current trend is a human in loop. The results of algorithms could be viewed as assisting tools to physicians to improve their productivity.
- Interpretable AI: When one includes humans (health professionals) in the loop, the question arises of whether the professionals would be comfortable with AI results, because they do not understand the minute details of the inference methodology. Deep learning models are typically exceedingly difficult to interpret.
- Nonavailability of data and labels: One must accept the fact that there are location-specific anthropometric, behavioral, and cultural differences. Collecting such a huge amount of data and labeling is a significant challenge.
- AI on the edge: AI on the edge devices provide low latency results. This also obviates the need for external network connectivity. Typically, the end user device could be a mobile phone. Battery life is a critical factor for such devices. Lean, yet efficient, computational models hold the key to success.

At this point in time, we can intersect three technologies: mobility, IoT, and AI. Chapter 3 will discuss the aspects related to data collection, signal conditioning, analysis, and ML with a focus on cardiovascular issues.

8. Devices: Surgical implants for cardiovascular monitoring and management

A medical device is defined as implantable if it is either partly or totally introduced surgically/medically into the human body and is intended to remain there after the procedure [32,33]. There are multiple devices for monitoring/treating cardiovascular conditions. Some of the devices are discussed in Chapter 2. Let us briefly discuss surgical implants.

There is a significant population in developed countries with implanted devices, enabling them to live a better life. Well-known examples are pacemakers, implantable cardioverter defibrillators (ICDs), and cardiac loop recorders. Pacemakers are used to manage bradycardia.

A pacemaker generates electrical pulses to keep the heart beating at the desired normal rate. These devices are implanted under the skin and are connected to the heart via leads. Leadless pacemakers also exist, which are placed inside the heart.

Implantable cardiac loop recorders are also placed under the skin. These devices record ECGs at preprogrammed intervals. These devices are typically for people whose normal ECG or Holter ECG (short term data) does not provide sufficient information about the condition of heart for diagnosis. Such conditions could indicate certain types of arrhythmias.

The wearable cardiovascular defibrillator is a noninvasive device which is prescribed to risky arrhythmic patients. This device detects arrhythmic events and generates high energy shock that depolarizes the myocardium and reinitiates the sinus rhythm [34]. There are computational models in place that can generate optimum electropathy in response to the type and degree of arrhythmia and may help in restoring sinus rhythm efficiently. Mobile-based ECG monitors can be used to monitor these patients [35].

The major challenge faced by the medical community is to analyze the huge amount of data produced by such an implantable device. If one takes the example of an implantable loop recorder, it can store up to 3 h of data at any instance of time. This data can be downloaded. Considering millions of such implanted devices, the total data is humongous and a human's ability to analyze or interpret this data is a nearly impossible task. At the same time, missing out any critical piece of data from this data pile could be extremely dangerous. Thus, the use of AI techniques is proving to be greatly beneficial in marking pathological conditions in this data and making it visible to healthcare professionals. This is the essence of this book. We will also discuss various methods of machine analysis of cardiac sensory data and abstracting meaningful insights.

9. Analytics for screening and diagnosis

Analyzing and interpreting the sensor data is the most important building block of connected health systems. There are many open frameworks suitable for this task. For the subsequent chapters on analysis and machine interpretation of various cardiovascular issues, a home-grown sensor data processing pipeline called the "automated feature discovery tool" was used. This platform automates machine-learning pipeline that transforms the time series data into 392 generic features. Examples of features that could be derived from raw ECG sample parameters include the RR interval standard deviation, and length of PQRS complex. From PCG (also known as a digital stethoscope) sound samples, a feature could be an average energy at 150 Hz signal. The tool then discovers a distinct feature set along with the construction of a learned model. A metaheuristic search is used for fast and efficient discovery of features from the pool of generated feature sets. This approach has outperformed state-of-the-art methods such as AutoML and related deep learning based networks [11].

10. Putting it all together

The book is divided in three sections. Section 1 (Chapters 1–3) describes aspects of NCDs with a focus on cardiovascular issues, sensing mechanisms, and an introduction to AI techniques. Section 2 (Chapters 4–7) describes how to process and analyze signals from these sensors and make meaning. Section 3 (Chapter 8) deals with future challenges.

Chapter 2 describes the evolution of unobtrusive sensing. The first formally known cardiovascular device is a stethoscope. It evolved from its earliest version of a wooden tube of specific dimensions, to its current binaural avatar. It further evolved into the form of a digital stethoscope in which the sound of heart pulses can be digitized, annotated, and processed. ECGs have also evolved from room-size machines to small credit-card-sized wearable devices [32,36]. This chapter will describe the basic building blocks of many cardiovascular sensing devices, how they can be used to predict cardiovascular issues, and a point of view on how they will evolve further.

Chapter 3 focuses on sensor signal analytics. Sensors play an especially vital role in converting the biosignals into a digital format. The raw output from sensors requires a lot of processing to make it usable. This chapter details aspects of signal preprocessing, processing, and postprocessing techniques. The preprocessing section will detail various techniques for sampling, filtering, and noise reduction for raw data. The processing section describes feature engineering, AI/ML techniques, and supervised/unsupervised learning. The postprocessing section talks about performance criterions such as accuracy, the F Score, and optimization parameters such as receiver opearting characteristic (ROC) curve. It shares the experience of handling unbalanced data from a practitioner's perspective.

Chapter 4 describes abnormal heart rhythms. The heart follows a rhythm, named the sinus rhythm. Rhythmicity is a function of harmonized operation of various functions of the heart related to generating electrical pulses by the SA node, and the opening and closing of valves. Any deviation from normal rhythm will generate abnormalities in the functioning of the heart. One example of such an abnormality with respect to heart rhythm is termed arrhythmia. This chapter discusses methods for collecting data and signal processing techniques for discovering arrhythmia using PPG and ECG data.

Chapter 5 discusses heart blockages, the reason for its occurrence, and digital sensing of heart blockages. It focuses on blockages in coronary arteries. Coronary arteries are blood vessels supplying blood to the heart muscles. If the arteries become blocked, this can lead to heart attack and even death. This chapter describes the mechanism of correlating data from ECG, PPG, and PCG to predict the blockage of coronary arteries. It describes the potential features in the above data that correlates with coronary blockages. It also talks about the real-world experience of fine-tuning the prediction models and features using data fusion.

Chapter 6 discusses hypertension and cardiac fatigue. Hypertension is one of the most widespread NCDs in the world. If untreated for a long time, it can result in complications such as cardiac arrest, diabetes, and blood clots in critical organs. This chapter describes the electrical simulation model of the heart using PPG data and other parameters. It also discusses the pulse transit time analysis from ECG and PPG for hypertension.

It is quite common for every muscle to become fatigued once it has been used for considerable time. However, heart muscles are the muscles that work nonstop. There are mechanisms to give microrests to heart muscles to recover. If heart muscles are not able to recover, they become fatigued. Because the symptoms are quite common, such as feeling out of breath, or feeling tired, they are often ignored by sufferers. However, these conditions could be dangerous in the long run. This chapter describes how to estimate cardiac fatigue by analyzing PPG and PCG data.

Chapter 7 discusses the other conditions and diseases that can lead to cardiac complications. One such disease is sleep apnea. Due to factors such as fibrous growth in the neck region or menopause, air flow to lungs can become constricted. This results in hypoxia and a person wakes up. This could be fatal if untreated. This chapter discusses relevant sensing mechanisms to find out the presence of this disease using frugal sensors.

Another such critical disease is chronic obstructive pulmonary disease (COPD) which is an inflammatory lung disease where air flow from lungs is constricted. It may be caused by factors such as smoking, exposure to harmful gases, and pollution. This chapter describes how a flow meter, along with other sensing elements, can be used to predict COPD.

Both diseases can result in impacts on the cardiac system as the oxygen supply pathway is affected and heart muscles may not receive sufficient oxygen or they have to work hard.

Chapter 8 takes the reader on a tour of the future possibilities. This includes disruptions in future sensing technologies such as flexible electronic-based wearables, using the human body as a communication medium, photoacoustic and hyperspectral sensing, radar sensing and computational imaging, nanobiosensing, and genomic analytics. It also covers advancements in AI such as AutoML, edge AI, neuromorphic computing, explainable AI, and related challenges such as privacy, transparency, trust, and security. Finally, the chapter outlines a hypothetical day in the life of a patient and doctor in 2030 that takes advantage of all the technology disruptions previously discussed.

11. Example of a real-world cardiovascular screening system

The following section describes how a real-world implementation of such algorithms can be done. The solution (Fig. 5) comprises of two parts. The first part is model development. This module takes annotated data as an input and churns out the models and algorithm as an output. This can be constructed using commercial off-the-shelf framework such as Pytorch/Tensorflow. It also could be an in-house solution optimized for healthcare data. Typically, this model follows an iterative approach where optimization

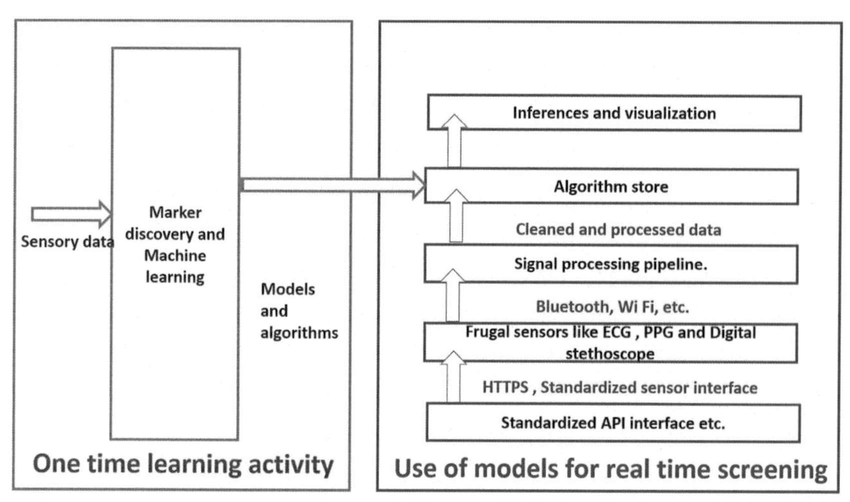

Fig. 5 A real life implementation of a screening system.

is carried out until the desired accuracy is reached. This is a computationally intensive activity and typically requires graphics processing units (GPUs).

Once the model/algorithm is developed, in order to make it usable, a PC-based or mobile or cloud-based application needs to be developed so that it can be used by the end user. Cloud-based applications can provide scalability.

An algorithm is a well-defined, structured process of solving a computational problem. As researchers work on multiple sets of data, newer and newer set of algorithms are developed. Typically, an algorithm takes sensor data as an input, processes it, and provides a computed output. An algorithm store is a place to host these algorithms in the platform [4]. One example could be an AF algorithm that takes raw ECG values as an input, computes parameters such as an RR interval, compares it with set of thresholds, and decides whether the sample under test has AF present in the signal or not. The challenge with algorithms is that, as researchers put in more effort, the algorithm gets better and better. Testing and hosting such upgraded versions of algorithms is a challenge often faced by researchers. Another aspect of algorithms is that cascading of algorithms can provide tremendous value. For example, say if an accelerometer value is given to a stepcounter algorithm, it can accurately count steps and associated parameters. However, if this algorithm is cascaded with, for example, "Gait" calculating algorithms, it can provide additional parameters such as stance, double leg support, and so on. Further the algorithm of detecting "pathological Gait" can give parameters such as hemiplegic or parkinsonian gait.

The customer-facing entity could be a cloud-based app with standardized application programming interface (API). The API will collect data from the sensing elements and perform some edge processing such as noise reduction and data compression. This data will then be sent to the algorithms store, housing various prevalidated algorithms. These algorithms would classify the data and give an output that can be presented to the end user. This architecture potentially can accommodate multiple algorithms. It could also be possible to have cascading algorithms for better insights. A similar architecture was used for crafting and hosting the models described in the later parts of this book.

References

[1] Business4, Business 4.0™: Digital Transformation Frameworks for Business, TCS, 2022. http://www.business4.tcs.com/. (last accessed 18 February 2022).
[2] Embracing Healthcare 4.0. https://cdn0.scrvt.com/39b415fb07de4d9656c7b516d8e2d907/1800000006533719/b4adc66d266b/Siemens_Healthineers_Paper_Embracing_Healthcare_4-0_1800000006533719.pdf.
[3] R. Jiao, S. Commuri, J. Panchal, J. Milisavljevic-Syed, J.K. Allen, F. Mistree, D. Schaefer, Design engineering in the age of Industry 4.0, J. Mech. Des. 143 (7) (2021), https://doi.org/10.1115/1.4051041.
[4] K. Bhavsar, H. Vishwakarma, B.R. Pawar, S. Shinde, S. Kimbahune, A. Ghose, A platform to enable algorithms as service model aimed at digital health service delivery, in: 2022 14th International Conference on Communication Systems & NETworkS (COMSNETS), 2022, https://doi.org/10.1109/comsnets53615.2022.9668453.
[5] J. Andreu-Perez, D.R. Leff, H.M. Ip, G.-Z. Yang, From wearable sensors to smart implants—toward pervasive and personalized healthcare, IEEE Trans. Biomed. Eng. 62 (12) (2015) 2750–2762, https://doi.org/10.1109/tbme.2015.2422751.
[6] Kardia, EKG Anywhere, Anytime, AliveCor, 2022. https://www.kardia.com/. (last accessed 18 February 2022).
[7] D.W. Roberts, The Commission on Chronic Illness, Public health reports (Washington, D.C.: 1896), U.S. National Library of Medicine, March 1954. https://www.ncbi.nlm.nih.gov/pmc/articles/PMC2024201/. (last accessed 18 February 2022).
[8] About Chronic Diseases, Centers for Disease Control and Prevention, April 28, 2021. https://www.cdc.gov/chronicdisease/about/index.htm. (last accessed 18 February 2022).
[9] NCBI, Roles of Biomarkers in Clinical Research, National Center for Biotechnology Information, 2022. https://www.ncbi.nlm.nih.gov/pmc/articles/PMC3078627/pdf/nihms259967.pdf. (last accessed 18 February 2022).
[10] WHO, Non Communicable Diseases, World Health Organization, 2022. https://www.who.int/news-room/fact-sheets/detail/noncommunicable-diseases. (last accessed 18 February 2022).
[11] A.S. Vaughan, R.C. Woodruff, C.M. Shay, F. Loustalot, M. Casper, Progress toward achieving national targets for reducing coronary heart disease and stroke mortality: a county-level perspective, J. Am. Heart Assoc. 10 (4) (2021), https://doi.org/10.1161/jaha.120.019562.

[12] M.E. Raichle, D.A. Gusnard, Appraising the brain's energy budget, Proc. Natl. Acad. Sci. U. S. A. 99 (16) (2002) 10237–10239, https://doi.org/10.1073/pnas.172399499.
[13] W. Wu, J. He, X. Shao, Incidence and mortality trend of congenital heart disease at the global, regional, and national level, 1990-2017, Medicine (2020). Wolters Kluwer Health https://www.ncbi.nlm.nih.gov/pmc/articles/PMC7306355/. (last accessed 18 February 2022).
[14] J. Stone, P. Hangge, H. Albadawi, A. Wallace, S. Fadi, M. Grace Knuttien, S. Naidu, R. Oklu, Deep vein thrombosis: pathogenesis, diagnosis, and medical management, Cardiovasc. Diagn. Ther. (2017). AME Publishing Company https://www.ncbi.nlm.nih.gov/pmc/articles/PMC5778510/.
[15] D.S. Park, G.I. Fishman, Development and function of the cardiac conduction system in health and disease, J. Cardiovasc. Dev. Dis. (2017). MDPI https://www.ncbi.nlm.nih.gov/pmc/articles/PMC5663314/.
[16] K. Maganti, V.H. Rigolin, M.E. Sarano, R.O. Bonow, Valvular heart disease: diagnosis and management, Mayo Clin. Proc. (2010). Mayo Foundation for Medical Education and Research https://www.ncbi.nlm.nih.gov/pmc/articles/PMC2861980/.
[17] C. Antzelevitch, A. Burashnikov, Overview of basic mechanisms of cardiac arrhythmia, Card. Electrophysiol. Clin. 3 (1) (2011) 23–45, https://doi.org/10.1016/j.ccep.2010.10.012.
[18] L. Staerk, J.A. Sherer, D. Ko, E.J. Benjamin, R.H. Helm, Atrial fibrillation, Circ. Res. 120 (9) (2017) 1501–1517, https://doi.org/10.1161/circresaha.117.309732.
[19] F. Sanchis-Gomar, C. Perez-Quilis, R. Leischik, A. Lucia, Epidemiology of coronary heart disease and acute coronary syndrome, Ann. Transl. Med. 4 (13) (2016) 256, https://doi.org/10.21037/atm.2016.06.33.
[20] W.-R. Tang, C.-Y. Yu, S.-J. Yeh, Fatigue and its related factors in patients with chronic heart failure, J. Clin. Nurs. 19 (1–2) (2010) 69–78, https://doi.org/10.1111/j.1365-2702.2009.02959.x.
[21] J.E. Ip, Wearable devices for cardiac rhythm diagnosis and management, JAMA Netw. (2019). https://jamanetwork.com/journals/jama/article-abstract/2721089.
[22] K. Rajakariar, A.N. Koshy, J.K. Sajeev, S. Nair, L. Roberts, A.W. Teh, Modified positioning of a smartphone based single-lead electrocardiogram device improves detection of atrial flutter, J. Electrocardiol. (2018). Churchill Livingstone https://www.sciencedirect.com/science/article/abs/pii/S0022073618303509.
[23] A.D. Dimarco, Diagnostic utility of real-time smartphone ECG in the initial investigation of palpitations, Br. J. Cardiol. (2018). https://bjcardio.co.uk/2018/03/diagnostic-utility-of-real-time-smartphone-ecg-in-the-initial-investigation-of-palpitations/.
[24] R. Dhingra, R.S. Vasan, Biomarkers in cardiovascular disease: statistical assessment and section on key novel heart failure biomarkers, Trends Cardiovasc. Med. 27 (2) (2017) 123–133, https://doi.org/10.1016/j.tcm.2016.07.005.
[25] WHO, More than 700 Million People with Untreated Hypertension, World Health Organization, 2022. https://www.who.int/news/item/25-08-2021-more-than-700-million-people-with-untreated-hypertension. (last accessed 18 February 2022).
[26] J.P. Kalehoff, S. Oparil, The story of the silent killer, Curr. Hypertens. Rep. 22 (9) (2020), https://doi.org/10.1007/s11906-020-01077-7.
[27] M. Mogi, Y. Higashi, K. Bokuda, A. Ichihara, D. Nagata, A. Tanaka, K. Node, et al., Annual reports on hypertension research 2020, Hypertens. Res. 45 (1) (2021) 15–31, https://doi.org/10.1038/s41440-021-00766-3.
[28] L.F. Drager, R.D. McEvoy, F. Barbe, G. Lorenzi-Filho, S. Redline, Sleep apnea and cardiovascular disease, Circulation 136 (19) (2017) 1840–1850, https://doi.org/10.1161/circulationaha.117.029400.
[29] W. Lee, S. Nagubadi, M.H. Kryger, B. Mokhlesi, Epidemiology of obstructive sleep apnea: a population-based perspective, Expert Rev. Respir. Med. 2 (3) (2008) 349–364, https://doi.org/10.1586/17476348.2.3.349.

[30] K. Park, Park's Textbook of Preventive and Social Medicine, Bhanot Publishers, India, 2017.
[31] A. Pal, Using AI on IoT Sensor Data—For Predicting Health of Man and Machine, 2022. https://site.ieee.org/indiacouncil/files/2019/12/p92-p94.pdf. (last accessed 18 February 2022).
[32] Y.-H. Joung, Development of implantable medical devices: from an engineering perspective, Int. Neurourol. J. 17 (3) (2013) 98, https://doi.org/10.5213/inj.2013.17.3.98.
[33] N.K. Mandsberg, J.F. Christfort, K. Kamguyan, A. Boisen, S.K. Srivastava, Orally ingestible medical devices for gut engineering, Adv. Drug Deliv. Rev. 165–166 (2020) 142–154, https://doi.org/10.1016/j.addr.2020.05.004.
[34] O. Mazumder, R. Banerjee, D. Roy, A. Mukherjee, A. Ghose, S. Khandelwal, A. Sinha, Computational model for therapy optimization of wearable cardioverter defibrillator: shockable rhythm detection and optimal electrotherapy, Front. Physiol. 12 (2021), https://doi.org/10.3389/fphys.2021.787180.
[35] C. Kropp, J. Ellis, R. Nekkanti, S. Sears, Monitoring patients with implantable cardioverter defibrillators using mobile phone electrocardiogra case study, JMIR Cardio (2018). Department of Psychology, and Corresponding Author: Caley Kropp. JMIR Publications Inc., Toronto, Canada. (Accessed 9 February 2022) https://cardio.jmir.org/2018/1/e5/.
[36] L. Lu, J. Zhang, Y. Xie, F. Gao, S. Xu, X. Wu, Z. Ye, Wearable health devices in health care: narrative systematic review, JMIR mHealth uHealth 8 (11) (2020), https://doi.org/10.2196/18907.

Further reading

S. Paul, M. Riffat, A. Yasir, M.N. Mahim, B.Y. Sharnali, I.T. Naheen, A. Rahman, A. Kulkarni, et al., J. Sens. Actuator Netw. 10 (3) (2021) 43, https://doi.org/10.3390/jsan10030043.
C. Thiene, G. Frescura, Anatomical and pathophysiological classification of congenital heart disease, Cardiovasc. Pathol. (2010). U.S. National Library of Medicine. (Accessed 9 February 2022) https://pubmed.ncbi.nlm.nih.gov/20466566/.

CHAPTER 2

Proliferation of a new generation of sensors: Smartphones and wearables

1. Introduction

The development of a sensing ability is a very important milestone in the history of humanity. The sensory system provides information that places the individual in relation to the environment [1]. There are five basic senses. These are sight (vision), taste (gustatory), smell (olfactory), touch (tactile), and hearing (auditory). Each corresponding sensing organ sends information to the brain and the brain makes meaning out of that information. There are a few other tertiary senses such as movement and balance (vestibular). Touch is the first sense that is developed when a baby is in the womb from just 8 weeks old. Other senses develop as the baby grows. Since Aristotle, it has been believed that there is a hierarchy of senses; vision as the most dominant, followed by smell, touch, taste, and sound. Sensing enables us to perceive things that are within and out of physical reach of the human body. These senses are vital for surviving and prospering.

For diagnosing diseases, sensing physiological and psychological parameters of a patient by medical professionals is vital. A diagnosis needs some type of observation window or quantification of body functions. Until the early 19th century, doctors looked at a small amount of physical/physiological aspects as a window to body functions. Some of the representative things were urine color, tongue composition, and skin pallor. Ancient doctors used to put patient saliva onto burning coal. The sweet smell of saliva, representing acetone, would point to the probability of the patient suffering from diabetes [2]. The fishy smell of ammonia would point to kidney-related issues. Alternatively, consider a simple stethoscope. It requires a significant amount of training and sharp hearing capabilities to identify heart diseases by just listening to the heart sounds. The placement of the stethoscope at the right location is also important to hear the right sounds.

New Frontiers of Cardiovascular Screening using Unobtrusive Sensors, AI, and IoT
https://doi.org/10.1016/B978-0-12-824499-9.00002-7

In today's world, electronic sensors in various forms provide the capability to sense medical parameters and provide inputs to healthcare professionals for correct diagnoses. There are multiple types of sensors. These sensors could be invasive sensors or noninvasive sensors.

The Cambridge English dictionary defines a sensor as "a device that is used to record that something is present or that there are changes in something." It also defines term "unobtrusive" as "Non noticeable – Seeming to fit well with the things around". A new generation of frugal and nonintrusive sensors have the potential to screen for various healthcare issues including cardiovascular diseases (CVDs).

In this chapter we will discuss the history and technical details of sensors that can provide digital data for various heart conditions. The focus will be on frugal and easy to use sensors. These sensors will be discussed in the context of control, physical, electrical, and fluidic functionality of the heart.

2. Unobtrusive digital sensing

The human lifespan is increasing as better medical treatments are evolving. Since 1840, life expectancy has increased at the rate of almost 2.5 years per decade [3]. The reasons could be a better understanding of the aging process, better treatment, and an overall increased focus on health and wellness. The cost of healthcare is also increasing proportionately. Particularly in developing or underdeveloped countries, scalability and last-mile reach of medical facilities is still an unresolved challenge. The doctor-to-patient ratio is skewed. Thus, offering health care services at a reasonable cost and scale is a great challenge. The solution to this challenge needs out-of-the box thinking.

As at June 2018, noncommunicable diseases (NCDs) have killed 41 million people per a year, equivalent to 71% of the deaths globally. Cardiovascular diseases account for 17.9 million annual deaths. The World Health Organization (WHO) has set a target of reducing 25% of deaths attributable to NCDs by 2025 [4]. Early detection and response could bring down health care expenses significantly. WHO has advised the use of digital technologies to manage this challenge.

In this book, we focus specifically on the use of frugal sensors and artificial intelligence (AI) technologies for screening of CVDs. Although sophisticated devices such as computed tomography (CT) scans, ultrasound, myocardial perfusion imaging, and methods such as coronary catheter angiography are available for cardiovascular diagnosis, their scalability is limited. A significant amount of research work is being undertaken on using commercial off-the-shelf sensors and machine learning (ML) techniques to develop a frugal CVD screening solution.

Smartphones are evolving at high speed and the adoption of smartphones exists at the grassroots level. In early 2020, worldwide, there were approximately 3.5 billion smartphones. That translates to over 44% of world's population being in possession of one. It is estimated that by 2025 there will be more than 18 billion mobile phones [5]. The smartphone is emerging to be the best last-mile connected device with reasonably good processing capability. Smartphones are equipped with a good number of sensors that can be leveraged to sense health parameters. The World Wide Web Consortium (W3) is an international community that develops open standards to ensure long-term growth of the web. Mobile is becoming the most preferred mechanism to access the Internet. W3 has formally enumerated the following sensors to be part of a mobile phone: ambient light, accelerometer, gravity, gyroscope, magnetometer, orientation, geolocation, and proximity [6]. Although not formally defined, there are implicit sensors such as a camera and microphone. External sensing devices such as electrocardiogram (ECG)/photoplethysmogram (PPG) glucometers can be connected to a phone and can provide significant value for health-related conditions. Some recent advances have also looked at using reflected Wi-Fi signals or Bluetooth signals as potential biosensors. Fig. 1 depicts a summary of sensors within and around mobile phones and smartwatches.

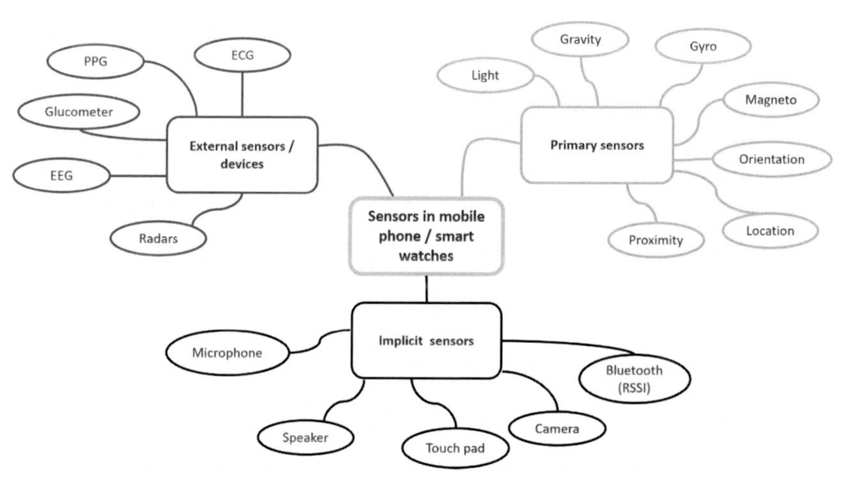

Fig. 1 Sensors in mobile phone.

Either as standalone modules, or a combination of these sensors, can provide a good framework for healthcare sensing. There has been significant work undertaken on leveraging the internal and interfaced sensors for healthcare screening applications. A camera in a mobile phone can be used as a sensing element for detecting SpO_2, heart rate [7], and trending blood pressure [8,9]. The microphone could be used to determine lung capacity [10,11]. Some work is being done on using a mobile touchpad as a sensing device [12]. The accelerometer provides the sensing capability for activity detection [13], calories spent, and so on. Mobile chips capable of running complex AI algorithms are evolving. A reasonably complex ML algorithm could be processed on the phone itself and newer insights could be discovered from this data [14].

Edge computing is an evolving area where the ML models are deployed on edge devices [15]. An edge device is any piece of hardware that controls data flow at the boundary between two networks. Traditional ML techniques are complex and require a large amount of computing resources. Typically edge devices are a small size with a limited amount of memory, computing power, and battery. Hence, running ML models on an edge device is tricky. TinyML aims to optimize these ML models to enable them to run on small footprint devices [16].

The case is the same when one looks at biosensing devices. As the miniaturization of technology evolves, devices have taken the form of wearables and patchables. Smartwatches with many sensors such as an accelerometer, PPG, and even ECGs started evolving. These devices are compact, always connected, and have edge-processing capabilities. Some of the sensors can be implanted in the body. One example is implantable loop recorders (ILRs) that are implanted under the skin and measure the ECG at preprogrammed intervals [17].

Sensors are now being embedded in clothes, hats, masks, and many items that people can wear [18,19]. Recently some devices have taken the form of patchable devices, i.e., can be patched on the skin. A typical example is a cardiac monitoring device with ECG, PPG, and Bluetooth capability [20,21]. These devices can be pasted on the chest and create data for 2–3 days. Some of the sensors can be swallowed [22] and they provide an inside view of the human body. These sensors are typically in the shape of a pill. They have a camera and other sensors. When swallowed, they provide imaging of the digestive path through wireless data. A new class of sensors called "earable" [23] is also evolving. It is based on the premise that the ear canal could be a very good place to receive accurate biosignals such as ECG, temperature, and SpO_2.

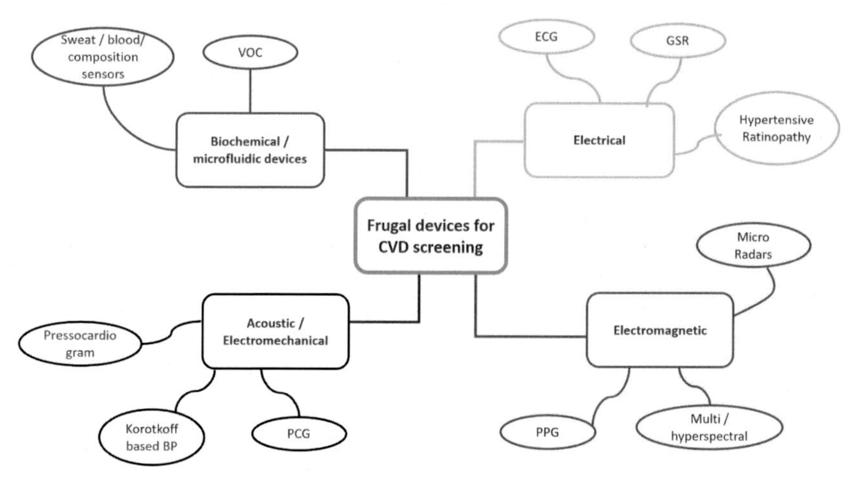

Fig. 2 Sensors for CVD screening.

Sensors enable longitudinal, objective, and accurate monitoring of physiological symptoms or conditions. This longitudinal data can provide vital clues to doctors for fine-tuning of treatment. This data can also be used for community research.

There are multiple sensing mechanisms for CVDs. These sensors can be categorized into the major classes that are discussed below. Fig. 2 depicts the classification and type of sensors in each category.

3. Electric or electromagnetic sensing

This class of sensors looks at the electromagnetic potential on the body surface or interaction with electromagnetic waves as the primary sensing media. A brief description of the devices/sensing mechanism is given below. Some of the devices are described in greater detail later in the chapter.

1. ECG: An ECG measures the electrical activity of the heart using electrodes positioned at strategic locations. ECGs typically comprise of 12 logical leads that provide a 360-degree view of electrical activities of the heart. The graph represents the time series of voltage values [24]. See later in the chapter for more details.
2. PPG: A PPG, measures volumetric changes of blood in capillaries and SpO_2 using light with specific wavelengths. It uses LEDs with specific frequencies to illuminate the skin and measure the reflected light [25]. See later in the chapter for more details.
3. EEG: Activities of the brain can be measured by using an electroencephalogram (EEG). An EEG measures the surface potential at various

regions on the skull and visualizes it for interpretations. Although it is not possible to measure the electrical potential generated by every neuron, the potential generated by a group of neurons can be measured. The potential is measured by placing electrodes on the scalp at predefined positions. There are two types of electrodes. Wet electrodes need gel, whereas dry electrodes do not need any adhesive. Each type of electrode has its advantages as well as limitations [26]. The 10–20 system is an internationally accepted method to apply electrodes on the scalp. It can also be used to identify a person [27], as well as detect emotions [28]. There is a significant amount of collaboration between the heart and the brain. Heart rate and its variations are controlled by changes in autonomic activities of sympathetic and parasympathetic nervous system [29]. This relationship is termed neurocardiology.

4. Radars: Radars are traditionally used for the detection of moving objects using electromagnetic waves. With recent advances in the miniaturization of radar technology, very small and cheap Doppler radars are available. A lot of research work is being undertaken on calculating the heart rate/breathing rate of a subject just by analyzing the reflected waves. Significant work has been done in identifying the blood flow patterns as well [30]. A smartspeaker can be used to monitor heart rate and rhythm in a contactless way [31]. The speaker emits an inaudible sound in the range of 18–22 kHz. The microphone captures the reflected sound waves from a human chest in the vicinity. It is analogous to SONAR. The reflected sound waves can be analyzed to find out parameters such as breathing rate, heart rate, and so on. It is based on the fact that when you are breathing, the chest compresses and expands. This delta change induces certain changes in the reflected sound waves. Although this approach has certain limitations, such as the person must be close to the speaker, and should not move, this provides a very simple yet effective way to measure parameters such as heart rate or breathing rate in a noncontact yet longitudinal way. This could prove very beneficial to people who are quarantined, or bedridden, where medical support is not available.
5. Hyperspectral sensing: The human eye can perceive wavelengths in the range of ~400–700 nm. Thus, information entropy is limited. Hyperspectral imaging enables us to take a series of images at a predefined frequency band over a much broader spectrum. The typical spectrum range is 300–2500 nm and the image resolution band is 20–30 nm. A set of images is termed a hypercube. A hypercube is a three-dimensional

database with two spatial and one spectral dimensions. To give an example, a normal human eye may not be able to differentiate between finely powdered sugar and table salt. However hyperspectral imaging (HSI) can definitely recognize the difference because the mean reflectance for these two items is different. Hyperspectral cameras are traditionally used in satellites for remote sensing and imaging. With miniaturization, it is being used in the medical field as well. It is used for disease diagnosis and surgery. Based on imaging, HSI provides diagnostic information about composition, morphology, and physiology of a cell or group of cells. It is believed that the light absorption and fluorescence profile changes for diseased tissues, therefore the reflected light from these tissues carry quantitative and qualitative diagnostic information about their pathology [32]. For example, hyperspectral imaging can reveal the composition of an atrial block [33] and quantify the amount of calcium or other chemicals. This could help with appropriate corrective actions. The technology trend indicates that hyperspectral sensing could soon be available in frugal and portable devices such as a mobile phone [34,35].

3.1 Acoustic and mechanical sensing

1. PCG: A phonocardiogram, also known as a digital stethoscope, converts the body sounds, especially heart sounds, to a digital format. Typically, it is a combination of conventional stethoscope with embedded electronics. Digital signals can be easily filtered, amplified, and visualized. AI and ML are enabling deeper understanding and interpretation of the heart sounds for identifying abnormalities. See later in the chapter for more details.
2. Blood pressure sensing: The main function of the heart is to pump the blood in the body at the correct pressure and rhythm. Blood pressure is measured in mm of mercury (mmHg). The conventional way of measuring blood pressure is by using a mercury sphygmomanometer. It is based on the Korotkoff sounds. In this method, a cuff is pressurized and the pressure is slowly released. At a particular pressure value, doctors can listen to the emergence of the Korotkoff sounds using the stethoscope. When pressure is further reduced, the Korotkoff sound weans off and at a particular point, it stops completely. The first pressure value when the Korotkoff sound is heard is termed the systolic pressure and the second value when the Korotkoff sound diminishes is termed the diastolic pressure. Recording the heart's systolic and diastolic pressure

requires a sufficient amount of training and practice. One needs to have a keen ear and sharp senses to record values [35]. A small error may result in an incorrect diagnosis. Digital blood pressure meters use electronically-controlled inflatable cuffs and pressure sensors to automatically measure the blood pressure. Commercially-available digital blood pressure meters are easy to use. However, their accuracy depends on various factors such as adherence to blood pressure checking protocol, the quality of the sensors, and the age of the device. Typically, they provide accuracy between $\pm 10\%$ [36]. Still, a mercury-based sphygmomanometer is considered the gold standard. There is a significant amount of research developing a simple mechanism such as using PPG analysis to measure blood pressure [37], or combination of ECG and PPG to estimate blood pressure values. [38]. However, accuracy and consistency is still an unresolved issue.

3.2 Biochemical sensing

1. Biomarkers in blood and sweat: Recent advances in microfluidics are enabling these types of devices that provide rapid and accurate detection of some of the biomarkers for cardiovascular issues. One such device uses a microfluidic chip to detect the concentration of cardiac troponins in very small concentrations accurately at the nanogram level [39,40]. It is a biomarker for acute myocardial damage.
2. Volatile organic compounds: Volatile organic compounds (VOCs) are organic compounds that evaporate at room temperature. Living entities release VOCs as a byproduct of their primary or secondary metabolisms. Similarly, there are VOCs present in the atmosphere produced by plants and bacteria. These VOCs can be absorbed through the skin. VOCs also are odor-producing compounds. There are over 3000 VOCs detected in human breath. However normally there are approximately 200+ VOCs present in the breath. Although VOCs are not harmful for short term exposure, long term exposure may lead to eye/throat irritation, headaches, and may damage kidneys and the nervous system. One major mechanism of communication between plants is via VOCs. An experiment by Babikova confirms this hypothesis [41]. The chemical analysis of breath can provide vital cues about the overall health of the person. Many clinical trials have been carried out to correlate the presence of specific VOCs and disease. The trials were in the area of detecting respiratory disorders, cardiovascular issues, diabetes, and digestion-related issues. Typical biomarkers (VOCs) were acetone for diabetes, methyl nicotinate

for tuberculosis, ammonia/alkylamines for renal function, and branched chain hydrocarbons for heart transplant rejections [41].

4. Sensing the heart: An engineer's perspective

The heart is a very important organ of the body. It has been studied extensively from biological and pathological perspectives. Here we present a different perspective of looking at the heart. The view presented below is from an engineer's perspective. How an engineer, rather than a doctor, would perceive the heart.

An engineer may view the heart as a controlled electromechanical fluidic system. Let us understand the heart from this perspective so that digital sensing of each of the aspects can be well understood and can be interpreted in the correct way. Fig. 3 depicts an engineer's view of the human heart.

4.1 Mechanical system

The heart comprises of four compartments and four valves. The upper two compartments are the atrium (right and left). The lower compartments are called ventricles (right and left). The right atrium receives deoxygenated blood from various body parts through the superior vena cava. Through the tricuspid valve, blood goes to the right ventricle. From the right ventricle, through the pulmonary valve, it is pumped to the lungs via the pulmonary artery. The left atrium receives oxygenated blood from the lungs through the pulmonary vein. This blood goes to the lower ventricle through the mitral valve. In the left ventricle, through the aortic valve, oxygenated blood is pumped to the body via the aorta. The opening and closing of valves could be viewed as a mechanical system.

Opening and closing of these valves and the gushing flow of blood generate sound and mechanical vibrations. These sounds could be digitized and processed for better insights into a heart's condition [42].

4.2 Circulatory system

The cardiovascular system, or in simple terms a circulatory system, is a conduit to circulate blood throughout the body [43]. Blood acts as a carrier for essential nutrients such as oxygen and glucose for every cell and takes back the byproducts such as carbon dioxide and other cellular waste for purification. This system comprises of arteries, veins, and capillaries. Capillaries act as a bridge between arteries and veins. As capillaries are nearer to the skin,

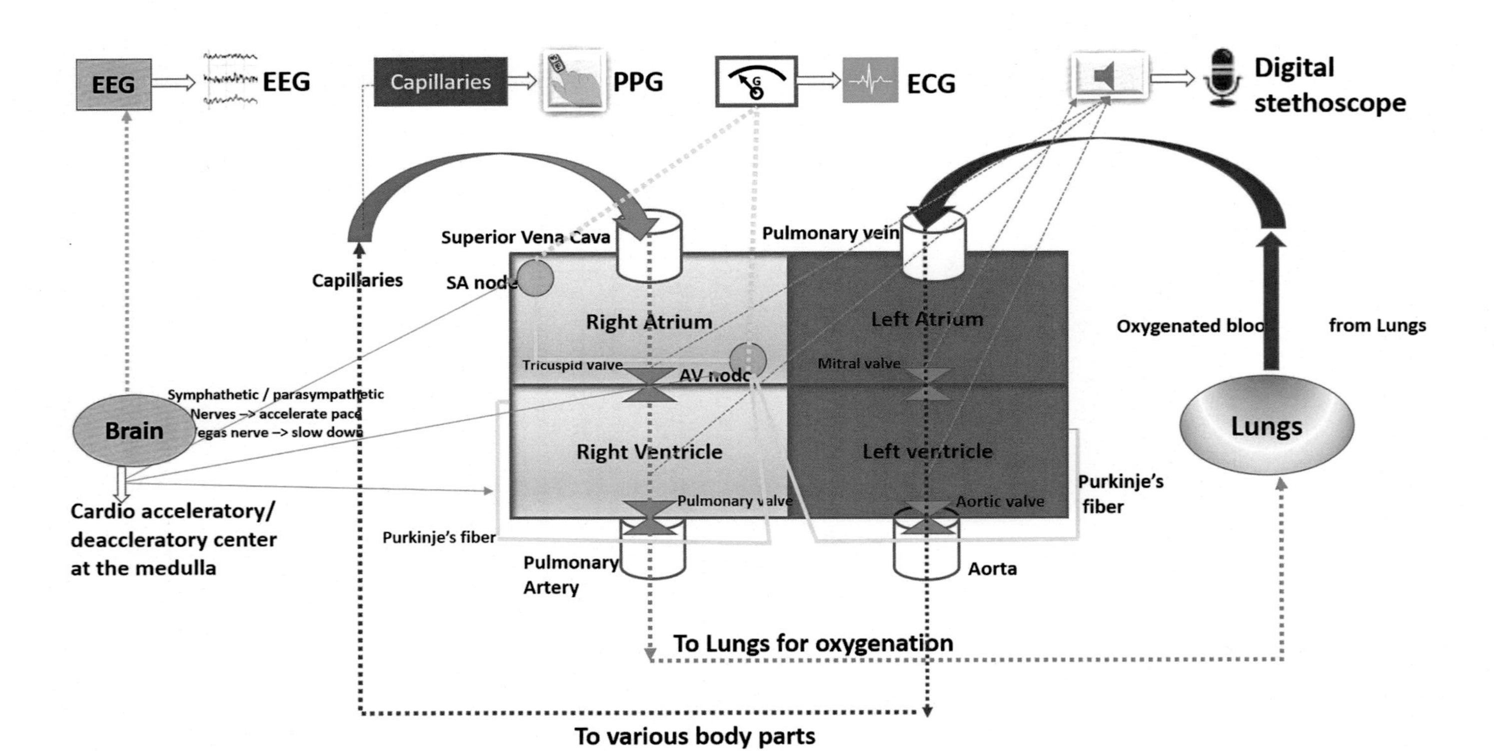

Fig. 3 Engineer's view of the heart.

it is easy to sense the blood flow volume and pattern by using simple pulse oximeters. The pattern arising out of a PPG wave can also provide vital cues on the condition of the heart [44].

4.3 Electrical system

The cardiac conduction system produces electrical impulses that move through the heart. These electrical signals are responsible for contractions of heart muscles and generate pressure to force blood through the circulatory system. It comprises of components such as the sinoatrial node (SA node), atrioventricular node (AV), bundle of HIS, and Purkinje's fibers [43].

Using ECG, these electrical signals could be measured at the peripheral skin level, processed and analyzed for insights of heart conditions including the electrical system.

4.4 Control system

Although the heart is more or less autonomous, it cannot be viewed in isolation without the brain. The brain produces signals such as the pacing up or down of the heart rate and connects to the heart through the sympathetic/parasympathetic nerves [43]. One example is the baroreflex control that maintains the correct blood pressure. However, for the sake of simplicity, only the controlled heart rate pacing is depicted in the diagram. EEG could provide vital clues on this aspect [45].

4.5 Systems coming together

Digitized signals from the above four systems could provide information regarding the condition of the heart. Although they can be viewed in isolation and give good insight, by fusing the data, we can receive a very detailed insight on the overall condition of the heart. The next chapters describe signal processing and insights from individual sensing and then fusing the data.

Now, with this background, let us dive deep into a few frugal sensing devices that can help when screening the condition of the heart.

5. Photoplethysmogram: pulse oximetry

Oxygen is a vital component for the human body to survive. Mother Nature provides oxygen abundantly in the air. The lungs absorb oxygen from the air and transfer it to the blood. The heart pumps this oxygenated blood throughout the body and the cells receive oxygen along with other nutrients. Conversely, the deoxygenated blood is pumped back into the lungs and is oxygenated. Hemoglobin plays a crucial role as an oxygen carrier. Without oxygen, all the cells in the human body will die. The deficiency of oxygen is termed hypoxia.

In the past few decades, to be able to measure oxygen saturation, a blood sample was required to perform tests such as arterial blood gas analysis. It is a complicated and time-consuming process. With the advent of digital technologies, a PPG or pulse oximeter is available that measures the blood flow volume and pattern in capillaries and the level of oxygen in the blood in a noninvasive yet simple way. The word photoplethysmogram evolved from two Greek words: Plathusmos, meaning to increase, and graphs, to write. It has become an important component of operation theaters and intensive care units (ICUs), along with other devices. It is generally safe to use pulse oximeters [44]. Although PPG sensing elements could be placed at locations such as the forehead or earlobe, the most popular location is the fingertip [46]. Oxygen is present in the blood in two forms: Dissolved, or bound to the hemoglobin. Hemoglobin can be functional or nonfunctional in terms of oxygen binding and transportation. Functional hemoglobin binds and transports oxygen. Hemoglobin bound with oxygen is termed oxyhemoglobin. When oxygen is detached from hemoglobin, it is termed deoxyhemoglobin. Nonfunctional hemoglobin is present as carboxyhemoglobin and methemoglobin. Carboxyhemoglobin is hemoglobin bound to carbon monoxide. Methemoglobin has ferric ions. PaO_2 is the partial pressure of oxygen dissolved in atrial blood. The percentage saturation of oxygen bound to hemoglobin in atrial blood is called the SaO_2 or SpO_2. There is a nonlinear relationship between SaO_2 and PaO_2.

A PPG typically comprises two light-emitting diodes, a photodetector, and signal processing circuit [44,47] (Fig. 4).

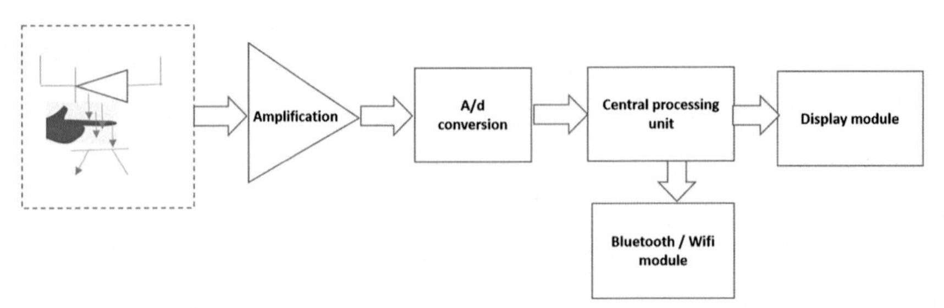

Fig. 4 PPG block diagram.

A PPG sensor works on the principle that oxygenated hemoglobin (HbO_2) and deoxygenated hemoglobin (Hb) absorbs the light in different ways. The absorption of infrared light for deoxyhemoglobin is higher at wavelengths from 600 to 800 nm, whereas the rate of absorption of oxyhemoglobin is higher at wavelengths from 800 to 1000 nm. Typically, one light emitting diode (LED) emits light at 660 nm (at which the rate of absorption of deoxyhemoglobin is greater than that of oxyhemoglobin). The other LED emits light in an infrared segment with a wavelength of 940 nm in which oxyhemoglobin absorbs more light than deoxyhemoglobin. Looking at the absorption values, the processor determines the value of oxy and deoxyhemoglobin. Photodiodes are strobed to record pulsative and nonpulsative flow. Processing is undertaken to arrive at an arterial blood flow pattern and O_2 saturation. The formula for calculating SpO_2 is:

$$\%\text{oxyhemoglobin } (SpO_2) = HbO_2/(HbO_2 + Hb).$$

A PPG signal comprises of two components:

- Pulsatile or AC component: This arises from the variations in blood volume by heart beats.
- Superimposed or DC component: A result of multiple factors such as respiration, sympathetic nervous system, and thermoregulation [44].

The output waveform has two peaks, namely systolic and diastolic peaks. Various features can be calculated by analyzing this waveform [37,48] (see Fig. 5). The derived features such as heart rate variability and peak to peak interval could reveal many factors such as heart condition, fitness level, atrial stiffness, and so on. The detailed signal processing will be covered in subsequent chapters.

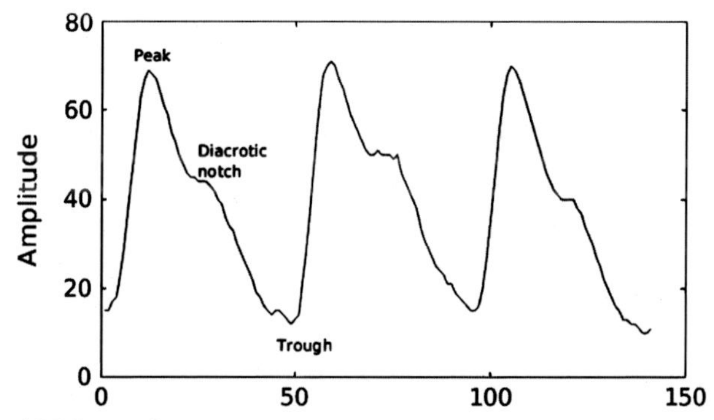

Fig. 5 Typical PPG waveform.

Whilst PPG can be collected from various parts of the body [44] (fingers, ear lobe, or forehead), the most popular choice is a finger, especially the middle finger of the right hand. The most common issue observed is inaccuracies due to incorrect placement and motion artifacts.

In the past few years, PPG has found a place in smartwatch fitness trackers as a consumer device. Many sports enthusiasts use it to monitor their heart rate oxygen level and adjust the exercise regime or running speed accordingly [44]. One needs to be aware of the fact that the readings from these wearables could be inaccurate [49]. Hence for medical applications/inferences, professional medical advice must be taken.

Further chapters in this book provide an indepth analysis of these features and their relationship with cardiac issues.

6. Phonocardiograms: digital stethoscope

When somebody visualizes a doctor, two things quickly come to mind. The first thing is a white coat, and the second is a stethoscope around the neck. The stethoscope has been there for approximately 200 years and not much has changed. Although it has some limitations, it is still a basic cardiac auscultation device used by most doctors worldwide. The Webster dictionary defines a stethoscope as "a medical instrument for detecting sounds produced in the body that are conveyed to the ears of the listener through rubber tubing connected with a piece placed upon the area to be examined" [4]. The word "stethoscope" is formed by joining two words: "stethos" meaning chest, and "scopein" meaning to examine. The stethoscope was invented by a French doctor, Rene Laennac, in 1826 when he was examining a young woman with a possible heart disease. He felt the need for a mediated auscultation device. Thus he crafted a customized wooden pipe to listen to heart sounds [50]. The earliest stethoscope was used for diagnosis of cardiac issues, however with the advances of digital technologies and availability of tools such as digital stethoscopes, ultrasounds, and CT scans, the stethoscope is now primarily used as a screening device. The use of a stethoscope has become an art rather than a science as it has many variables that need to be interpreted by just listening to the various sounds. The US FDA classifies manual stethoscopes under class I category (Regulation no 870.1875), whereas digital stethoscopes are covered under class II category (Regulation number 870.1875). The basis of classification is associated with the potential risk to patients with respect to use of that device. Devices are classified as class 1 when risk is minimal whereas devices are classified as class II when the risk to the patient is moderate to high.

6.1 History of heart sounds

In the pre-Hippocrates days, physicians performed auscultation of heart sounds by placing their ear directly on the patient's chest. This technique was called an "immediate auscultation." This auscultation technique was useful for listening to lung sounds as the lungs cover a large area, however it was not as good for cardiac examinations because it is difficult to locate the source of the heart sounds and there are multiple heart sounds with a minor difference. It needs an expert ear to identify the sound and associated issues correctly. Sounds were described in terms of the creak of new leather. Around 1400 BC, various scientists such as Leonardo da Vinci and others became interested in cardiac auscultation [50,51]. In 1628 William Harvey formally defined the heart sounds in his book, *De Motu Cordis*, that in each moment of the heart, when there is the delivery of a quantity of blood from the veins to the arteries, a pulse can be heard within the chest. He described two recurring pulses from the heart. In 1757 William Hunter first identified a murmur sound from the heart. Allan Burns detailed heart murmurs further. In the early 18th century, immediate auscultation was not very suitable for cardiac sounds because it did not meet the need of cultural beliefs and hygiene. Consequently Laennec used a rolled sheet of paper in the form of a cylinder as an auscultation device. He further experimented, improvised the device and came up with a design of 25 cm × 2.5 cm hollow wooden cylinder. Later it took the form of three detachable cylinders. This was the first instance of "mediated auscultation." He contributed significantly to the understanding and interpreting of cardiac auscultation sounds [51]. In the 19th century Arthur Leared created a binaural stethoscope and later Philips Cammann improvised the design. Today the Littmann brand is one of the most popular stethoscopes.

6.2 Heart sounds

Heart sounds are a result of acoustic waves generated by heart muscles, including the opening and closing of valves and resultant blood flow. There are two normal sounds in a healthy human being. They are "Lub—S1 or systolic tone" and "Dub—S2 or diastolic tone". There can also additional sounds including murmurs or gallop rhythms (S3 and S4). Fig. 6 depicts a typical waveform of a normal heart sound.

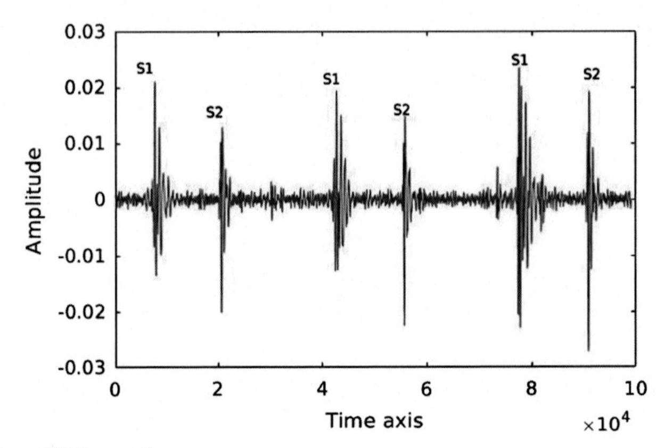

Fig. 6 Typical PCG waveform.

Heart sounds can be classified into normal and abnormal sounds. Typically, heart sounds are named as S(n) where n varies from 1 to 4. S1 and S2 are normal heart sounds. S1 is produced at the beginning of cardiac cycle (start of systole). It is short but a high-pitched sound. S2 is a second heart sound when systole ends. It is also a short yet high-pitched sound. Conventionally S1 is termed "Lub" and S2 is termed "Dub" sounds. S3 and S4 are classified as abnormal heart sounds. S3 is heard at an early diastole stage. It is a short yet low-pitched sound, hence one needs a keen ear to locate it. Similarly, S4 could occur at a late diastole. The presence of S3 or S4 indicates abnormality in the functioning of the heart. Apart from these sounds, there are other sounds that point to defects. These are sounds related to opening snaps, ejection sound, and murmurs (low pitch and high pitch). Murmurs could represent conditions such as mitral stenosis, mitral regurgitation, aortic stenosis, or mitral valve prolapse [52]. Heart murmurs are blowing or whooshing sounds caused by turbulent blood flow patterns. All murmurs are not pathogenic.

The average heartbeat cycle of an adult is ~0.8 s; an average of 0.3 s of systole and 0.5 s of diastole. The average ratio of S1 to S2 is 1:2.

6.3 Conventional stethoscope

A stethoscope (Fig. 7) comprises of the following parts:

- Chest piece: This is a tunable diaphragm that generates high or low frequencies based on pressure changes. It can be used from both sides depending on the auscultation needs. There are different types of chest pieces available catering to adults, pediatric, or fetal auscultations.

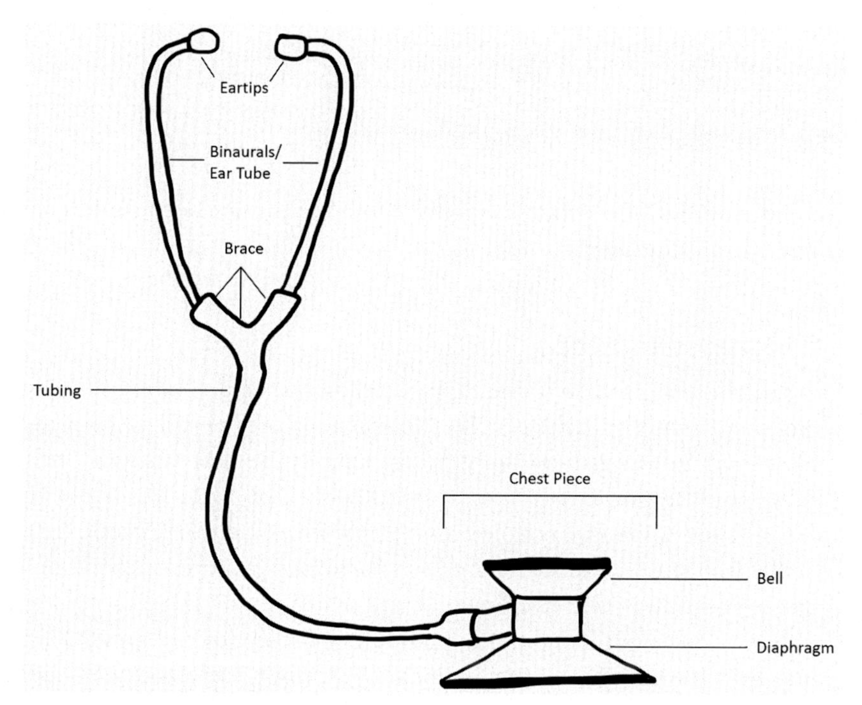

Fig. 7 Conventional stethoscope.

- Ste This connects the chest piece to flexible tubing.
- Flexible tubing: This carries sound waves generated by the chest piece toward earpieces.
- Binaural spring: This spring connects the flexible tubing to binaural steel tubes. They also enable adjusting the tension for binaural tubes so that the ear pieces fit snugly in the ear. This helps in listening to the sounds correctly as well reducing background noise.
- Binaural tubes and earpieces: A physician can listen to the sounds through two ear tips.

Cardiac auscultation using a stethoscope needs a keen ear and a deep understanding of the heart functions [52]. The normal frequency of pulmonary sound varies from 100 to 1000 Hz, however wheezing sound frequencies may vary from 100 to 5000 Hz.

With the advances of digital technologies in the 20th century, digital stethoscopes have started emerging in the market. A digital stethoscope converts cardiac/other sounds to electronic signals [42,53]. As these signals are electronic in nature, they can be easily filtered, amplified, and visualized. With the emergence of AI and machine learning, a significant body of

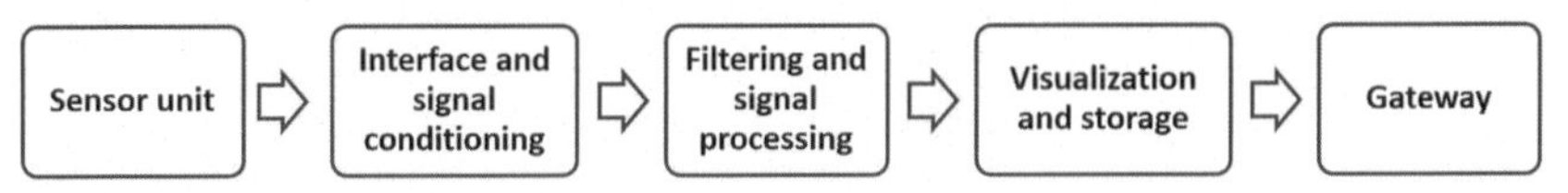

Fig. 8 Basic building blocks of digital stethoscope.

research work has been done in machine interpreting the heart sounds and identifying abnormalities.

6.4 Digital stethoscope

The basic building blocks of the digital stethoscope (Fig. 8) are:

1. Sensing module (transducer): This unit picks up the pressure changes and converts it into digital signals. Although initially piezoelectric transducers were popular, capacitive sensors have taken their place because they offer more fidelity. The main challenge is that the heart sounds are produced at lower frequencies. Hence it is more prone to pick up noises such as ambient sound or a power line hum.
2. Interface and signal conditioning module: This module collects the signals from the transducer and amplifies it. Post amplification, antialiasing is undertaken. This helps in reducing the artifacts in the signals. Afterwards, signal encoding is done.
3. Digital signal processing module: The major tasks of this are:
   - Sound denoising: Each signal contains many internal and external sounds. Examples of internal sounds are sounds generated by rubbing of the rubber tubes and movement artifacts. External sounds could be background noise, noise generated due to other body functions, and so on. There are various techniques to minimize these noises. One of the most popular ones is wavelet denoising based on discrete wavelet transform (DWT).
   - Heart sound segmentation: This module looks at the data and tries to find peaks in the audio signals. These peaks correspond to S1, S2, and other sounds. To achieve this, segmentation of the signal window is done. There are two approaches to segmentation: direct segmentation, where the envalogram based approach is very popular; and indirect segmentation, which looks at other signals such as ECG to try to segment the signal. Direct segmentation calculates the envelope size using a normalized Shannon's energy criterion. It detects the peaks in an envelope and establishes a correspondence between S1, S2, and other peaks.

4. Feature extraction and classification: In order to classify the peaks, as to whether they are normal or not, various ML techniques are used. Reasonably good, annotated data sets are available for benchmarking. Mel-frequency cepstral coefficients (MFCC) and discrete wavelet transform (DWT)-based features are widely used for classification of sounds.
5. Visualization module: This module provides a neatly formatted display of waveforms [47].
6. Gateway module: This module connects the device to the external networks. The connection channel could be through Bluetooth low energy (BLE), Wi-Fi, or data channels such as GSM/CDMA phone bands.

6.5 Future of the stethoscope

The stethoscope is rapidly becoming digitized. AI is being built into the stethoscope itself. Integrated devices such as a combined stethoscope and ECG are on their way. A few nonconventional usages of old technologies such as radar are also being investigated to recognize heart rate, breathing rate, and blood flow patterns. Having said that, widespread acceptance of these tools would take some time; in the meantime, the stethoscope will be still the basic auscultation tool [42,53].

7. ECG: Electrocardiograph

Cardiovascular issues have been known to mankind for a long time. However, its evidence-based pathology started in the late 18th century when the X-ray was invented. X-rays provide some information on the structure of the heart and major blood vessels. Understanding the pathology of the heart took a giant leap forward with the invention of a 3-lead ECG by Einthoven in 1901. This technology has further evolved and is currently a 12-lead ECG system. Fig. 9 briefly traces the history of the evolution of the ECG [54].

The ECG is probably one of the most important sensing mechanisms that provide a wealth of information about the condition of the heart. Hence

1842	1887	1901	1938	1942	1952
Matteucci Recorded electricity From frogs heart	Waller recorded Electrical activity From human heart	Enthoven built String galvanometer- 3 lead EKG machine	Wilson invented Precordial leads	Goldberg invented Unipolar leads	AHA standard 12 leads ECG

Fig. 9 Evolution of ECG.

it is prudent to understand the biological and electrical aspects that can influence an ECG.

7.1 Origination and pathways of electrical charges in the heart

The cardiac conduction system produces electrical impulses that move through the heart. These electrical signals are responsible for the contractions of heart muscle and generating pressure to force blood through the circulatory system. These signals travel across the surface of the body and can be measured. This system is autonomous and is not directly controlled by the brain. Fig. 10 depicts the electrical system of the heart.

The major functional units in this system are:

1. Sinoatrial node (SA node): This node is located in the wall of the right atrium. It is banana-shaped and is 10–30 mm in length and 5–7 mm in diameter. This node comprises cells that can produce electrical impulses. These impulses travel through an electrical conduction system on the walls of the heart, causing it to contract. Autonomic nerves of the peripheral nervous system regulate the pacing of the heart.
2. Atrioventricular node (AV): This node lies on the right side of atrium. It acts as a gateway of an electrical signal to ventricles. It checks the signal from the AV node. If it is acceptable, it acts as a gateway, otherwise it produces its own impulses. There is a delay of approximately 100 ms for an electrical signal to travel from the SA node to the AV node. This ensures the atria have ejected all the blood in the ventricles.
3. Bundle of HIS: The AV node passes electrical signals to a bundle of HIS. There is a left and right branch of HIS.

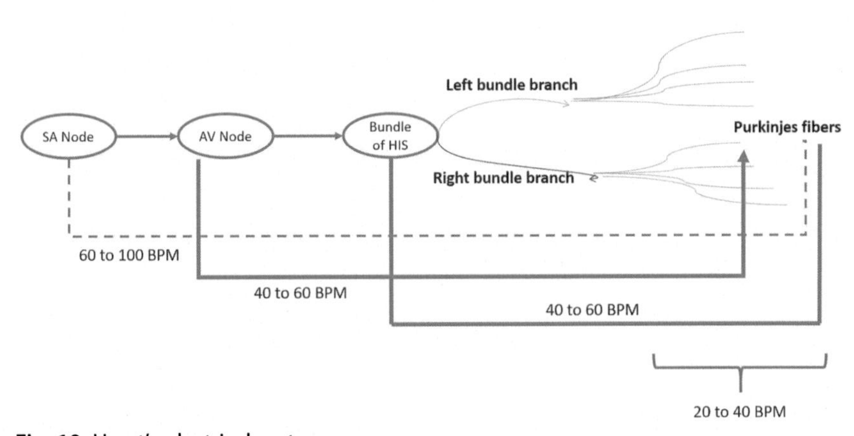

Fig. 10 Heart's electrical system.

4. Purkinje's fibers: This is the last component of the electrical system. It is responsible for distributing the electrical signal over ventricles and enables them to contract [43].

When cardiac signals are generated by the above mechanism, they are used for the functioning of the heart. However, they spread through adjacent tissues and a small portion comes to the surface of the body. By measuring this potential, one can get the macro view of the electrical activities of the heart. Electrical/mechanical and hydraulic systems work in tandem. Hence problems in one system are reflected in all the other systems in varying proportions. For example, if there is an electrical malfunction, the blood might not be completely drained, thus giving some features in the acoustic space.

The electrical activity can be recorded and plotted on an ECG. The voltages typically depend on how the electrodes are applied and the distance from the heart. Fig. 11 depicts a typical ECG waveform.

An ECG waveform comprises of the following repetitive segments:

1. P wave: This represents the start of the cardiac cycle. This portion of electrical activity is a result of atrial depolarization. The typical P wave is 0.12s. It looks like the upper half of a sinusoidal wave.
2. QRS complex: This represents the middle part of the cardiac cycle. This portion looks like an inverted V. This complex comprises of three separate waves: Q wave, R wave, and S wave. This complex is a resultant of ventricular depolarization. The typical time duration of this complex is 0.06s. Typical peak voltage is 3–4mv.
3. T wave: This represents the repolarization of ventricles. It also looks like a chopped portion of the sine wave. The typical duration of T wave is

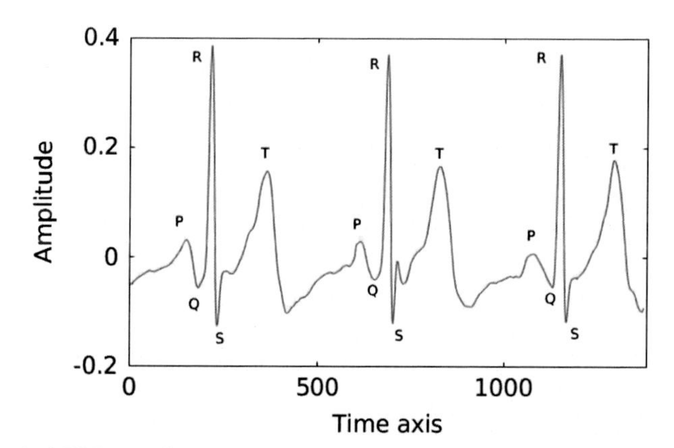

Fig. 11 Typical ECG waveform.

0.25–0.35 s. Atrial repolarization happens but it becomes shadowed by ventricular polarization.

4. ST segment: Connects the QRS complex and T wave. This represents depolarized ventricles.
5. J point: This represents the point where QRS complex finishes and ST segment begins.
6. PQ or PR interval: The time between the beginning of P wave and QRS complex, which represents the starting of excitation of atria and ventricles, is about 0.16 s. If the Q wave is absent, it is called a PR interval.
7. QT interval: This interval represents the time from the contraction of ventricles to the end of the T wave. This is typically 0.35 s.

The above parameters form the basis of digital signal processing in an ECG waveform.

7.2 12 leads and Einthoven's triangle

The correct placement of leads for recording electrical signals can provide immense insights into the current state of the heart. The principle for placing the electrodes is based on Einthoven's triangle [54].

There are a total of 12 leads that provide a detailed view of the electrical activity of the heart in 3D space. The electrodes are categorized into the following three categories.

Three bipolar limb leads

The ECG is recorded from two probes placed on different sides of the heart:

- Lead I: ECG probe connected to the right arm is treated as a negative terminal, whereas a probe connected to the left arm is treated as a positive terminal.
- Lead II: Probe connected to the right arm is taken as a negative terminal, whereas a lead connected to left leg is treated as a positive terminal.
- Lead III: Here the ECG probe connected to the left arm is treated as a negative terminal, and probe connected to the left leg is treated as a positive terminal.

Three augmented leads

- aVF lead: Lead placed between right arm/left arm as a negative lead, and left leg as positive lead.
- aVL lead: Lead placed between right arm left leg as negative lead, and left arm as a positive lead.
- aVR lead: Lead placed between left arm/left leg as negative lead, and right arm as the positive lead.

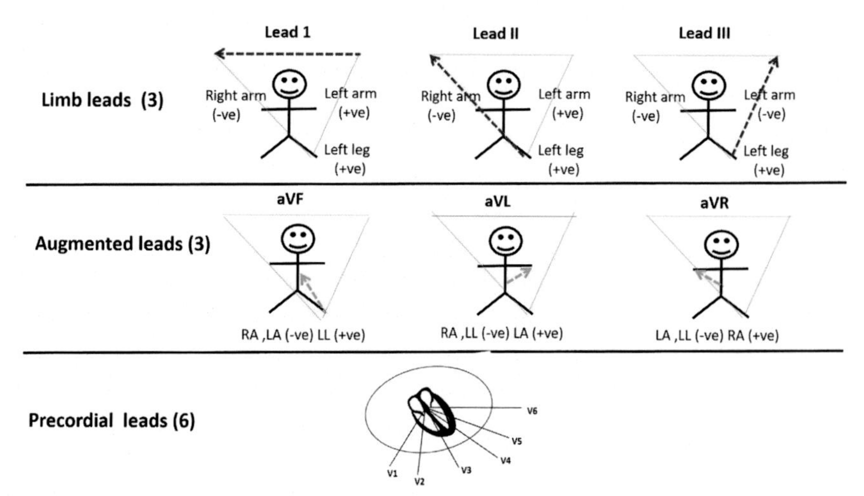

Fig. 12 Summary of placement of ECG leads.

Six precordial leads

For a normal person, the vector of depolarization, polarized from the right upper area (SA) to left lower area of the heart. It aligns with Lead II. It provides a 3D view of conduction in 3D space. It is perpendicular to the bipolar/augmented leads. There are six precordial leads placed at the strategic locations around the heart in the plane of 120 degrees from Lead I if viewed clockwise.

Fig. 12 summarizes the placement of leads.

The use of 12 leads provides a good insight of heart functions. Similar to other electrical/electronic systems, sometimes aging in hearts changes the values of resistance, capacitance, and system malfunctions. For some of these complicated issues, the insertion of electrodes and catheters in the heart is needed to measure action potential at various locations. The branch of electrophysiology deals with these types of issues. The ECG instrumentation required for this purpose is much more sophisticated than the 12-lead system.

7.3 Basic building blocks of an ECG machine

Fig. 13 depicts basic building blocks of a typical ECG machine [47,55]:

- Data acquisition module: This module interfaces the ECG sensing leads with the devices. The leads could be of capacitive or wireless type. This module multiplexes and amplifies the signal.
- Data preprocessing module: This module processes the data coming out from the data acquisition module. This module comprises appropriate

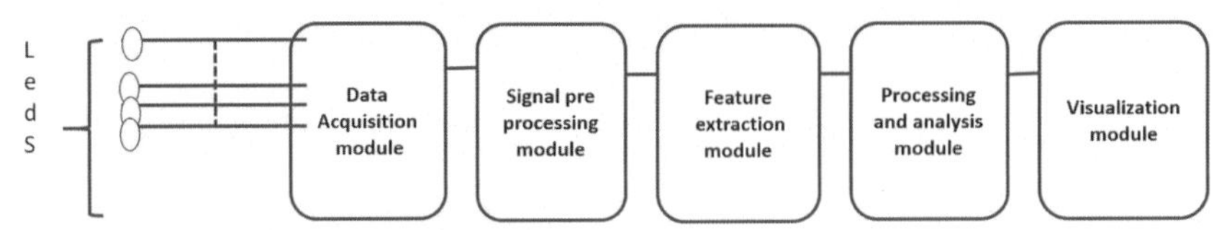

Fig. 13 ECG machine building blocks.

filters to reject out-of-band data. It cleans up data and normalizes it. It removes noise emanating from sources such as power lines, electrode contact noise, movement artifacts, and so on. Techniques such as wavelet transform and adaptive digital filters are used to process the data.

- Feature extraction module: This module processes the data and extracts the features from the data. This is the most important module as it provides interpretable information about the heart's parameters. The most popular feature extraction methods could be based on time-domain features such as the R-R interval or the S-T interval or frequency domain features such as FFT or power spectral density.
- Processing and analysis module: This module looks at the features and ascertains the quality of the signal. It could do peripheral calculations such as heartbeat computations.
- Visualization module: This module takes the processed ECG signals and displays it on the screen or can send it for printing. This is the most important module because the appropriate visual representation of signals can assist a cardiologist to assimilate the data and undertake correct decisions [56].

The ECG form factor has been transforming from hospital- or laboratory-based devices to consumer devices. It is becoming part of wearable devices as well. Many athletes regulate their schedule and plan their regimen according to readings based on an ECG, along with other sensors. The journey of ECG from the laboratory is currently to wearable devices, and it is soon moving toward patchable devices.

One class of ECG devices are an in vivo type. They are implantable loop recorders (ILRs) that are inserted in the human body and record the ECG for a considerable duration. These are needed when there is a requirement of recording ECG for an extended period of time due to issues such as frequent fainting or abnormal heartbeats. They are typically kept for 2–3 years and collect ECGs at specially programmed time intervals depending on the health issue requirement. The real challenge is to analyze this huge amount of data. The challenge becomes very hard with multiple thousands of devices that need to be analyzed. Consequently, the processing techniques described in this book could play a vital role.

8. Conclusion

In this chapter we have discussed how a human heart functions and how an engineer would view these functions of a heart through a digital window. This noted how to digitally sense the heart in a holistic way and provide pathways and generating meaningful insights from the sensor data. We also looked at interpreting the sensors' data. We discussed the history of the stethoscope, ECGs, and PPGs from their initial avatar to current versions, including the challenges with the possible evolution. In the next chapter, we discuss various processing and analyzing techniques for the sensors' data, and also produce an overview of techniques for extracting meaningful insights from this data.

References

[1] G.K. Bigley, Sensation, in: Clinical Methods: The History, Physical, and Laboratory Examinations, third ed., U.S. National Library of Medicine, January 1, 1990. https://www.ncbi.nlm.nih.gov/books/NBK390/.
[2] J. Rosenhek, Liquid Gold, September 1, 2005, Doctor's Review | Liquid Gold http://www.doctorsreview.com/history/sep05_history/.
[3] J.W. Vaupel, F. Villavicencio, M.-P. Bergeron-Boucher, Demographic perspectives on the rise of longevity, Proc. Natl. Acad. Sci. (2021). https://www.pnas.org/content/118/9/e2019536118.
[4] Stethoscope Definition & Meaning, Merriam-Webster, https://www.merriam-webster.com/dictionary/stethoscope.
[5] Published by S. O'Dea, and Sep 24 Number of Mobile Devices Worldwide 2020-2025, Statista, September 24, 2021. https://www.statista.com/statistics/245501/multiple-mobile-device-ownership-worldwide/.
[6] Generic Sensor API, W3C, December 15, 2021. https://www.w3.org/TR/generic-sensor/.
[7] A. Pal, A. Sinha, A.D. Choudhury, T. Chattopadyay, A. Visvanathan, A robust heart rate detection using smart-phone video, in: Proceedings of the 3rd ACM MobiHoc Workshop on Pervasive Wireless Healthcare—MobileHealth '13, 2013, https://doi.org/10.1145/2491148.2491156.
[8] A. Visvanathan, A. Sinha, A. Pal, Estimation of blood pressure levels from reflective photoplethysmograph using smart phones, in: 13th IEEE International Conference on BioInformatics and BioEngineering, 2013, https://doi.org/10.1109/bibe.2013.6701568.
[9] N. Bui, A. Nguyen, P. Nguyen, et al., Smartphone-based SPO2 measurement by exploiting wavelengths separation and chromophore compensation, ACM Trans. Sens. Netw. (2020). https://dl.acm.org/doi/abs/10.1145/3360725.
[10] A. Abushakra, M. Faezipour, Lung capacity estimation through acoustic signal of breath, in: 2012 IEEE 12th International Conference on Bioinformatics & Bioengineering (BIBE), 2012, https://doi.org/10.1109/bibe.2012.6399655.

[11] T. Thap, H. Chung, C. Jeong, K.-E. Hwang, H.-R. Kim, K.-H. Yoon, J. Lee, High-resolution time-frequency Spectrum-based lung function test from a smartphone microphone, Sensors 16 (8) (2016) 1305, https://doi.org/10.3390/s16081305.
[12] S. Kimbahune, S. Shinde, K. Bhavsar, A. Ghose, S. Khandelwal, A. Pal, Heart rate monitoring using capacitive touchscreen sensing, in: Proceedings of the Workshop on Body-Centric Computing Systems, BodySys'21, 2021, pp. 13–17, https://doi.org/10.1145/3469260.3469667.
[13] A.S. Sukor, A. Zakaria, N. Abdul Rahim, Activity recognition using accelerometer sensor and machine learning classifiers, in: 2018 IEEE 14th International Colloquium on Signal Processing & Its Applications (CSPA), 2018, https://doi.org/10.1109/cspa.2018.8368718.
[14] A. Prasad, A.K. Tyagi, M.M. Althobaiti, A. Almulihi, R.F. Mansour, A.M. Mahmoud, Human activity recognition using cell phone-based accelerometer and convolutional neural network, Appl. Sci. 11 (24) (2021) 12099, https://doi.org/10.3390/app112412099.
[15] Mobile AI: On-Device Ai: Qualcomm®, Qualcomm, February 10, 2022. https://www.qualcomm.com/products/smartphones/mobile-ai.
[16] P.P. Ray, A review on tinyml: state-of-the-art and prospects, J. King Saud Univ. Comput. Inf. Sci. (2021). Elsevier https://www.sciencedirect.com/science/article/pii/S1319157821003335.
[17] Implantable Loop Recorder, Johns Hopkins Medicine, https://www.hopkinsmedicine.org/health/treatment-tests-and-therapies/loop-recorder-implantation.
[18] D. Dias, J.P.S. Cunha, Wearable health devices—vital sign monitoring, systems and technologies, Sensors 18 (8) (2018) 2414, https://doi.org/10.3390/s18082414.
[19] J. Andreu-Perez, D.R. Leff, H.M. Ip, G.-Z. Yang, From wearable sensors to smart implants—toward pervasive and personalized healthcare, IEEE Trans. Biomed. Eng. 62 (12) (2015) 2750–2762, https://doi.org/10.1109/tbme.2015.2422751.
[20] D. Duncker, W.Y. Ding, S. Etheridge, P.A. Noseworthy, C. Veltmann, X. Yao, T.J. Bunch, D. Gupta, Smart wearables for cardiac monitoring—real-world use beyond atrial fibrillation, Sensors 21 (7) (2021) 2539, https://doi.org/10.3390/s21072539.
[21] IEEE Xplore, A Wireless Wearable Sensor Patch for the Real-Time Estimation of Continuous Beat-to-Beat Blood Pressure, IEEE Xplore, 2019 (Accessed 11 February 2022) https://ieeexplore.ieee.org/abstract/document/8857446.
[22] J.L. Toennies, G. Tortora, M. Simi, P. Valdastri, R.J. Webster, Swallowable medical devices for diagnosis and surgery: the state of the art, Proc. Inst. Mech. Eng. C J. Mech. Eng. Sci. 224 (7) (2010) 1397–1414, https://doi.org/10.1243/09544062jmes1879.
[23] K.B. Mikkelsen, Y.R. Tabar, S.L. Kappel, C.B. Christensen, H.O. Toft, M.C. Hemmsen, M.L. Rank, M. Otto, P. Kidmose, Accurate whole-night sleep monitoring with dry-contact ear-EEG, Sci. Rep. 9 (1) (2019), https://doi.org/10.1038/s41598-019-53115-3.
[24] M. Cadogan, History of the Electrocardiogram, Life in the Fast Lane, LITFL, February 3, 2022. https://litfl.com/history-of-the-electrocardiogram/.
[25] M. Ghamari, A review on wearable photoplethysmography sensors and their potential future applications in health care, Int. J. Biosens. Bioelectron. 4 (4) (2018), https://doi.org/10.15406/ijbsbe.2018.04.00125.
[26] M. O'Sullivan, A. Temko, A. Bocchino, C. O'Mahony, G. Boylan, E. Popovici, Analysis of a low-cost EEG monitoring system and dry electrodes toward clinical use in the neonatal ICU, Sensors 19 (11) (2019) 2637, https://doi.org/10.3390/s19112637.
[27] B. Hu, C. Mao, W. Campbell, P. Moore, L. Liu, G. Zhao, A pervasive EEG-based biometric system, in: Proceedings of 2011 International Workshop on Ubiquitous

Affective Awareness and Intelligent Interaction—UAAII '11, 2011, https://doi.org/10.1145/2030092.2030097.
[28] X. Yu, W. Qi, A user study of wearable EEG headset products for emotion analysis, in: Proceedings of the 2018 International Conference on Algorithms, Computing and Artificial Intelligence, 2018, https://doi.org/10.1145/3302425.3302445.
[29] E.E. Van der wall, The brain heart connection; a round trip, Neth. Heart J. 19 (2011) 269–270. https://link.springer.com/article/10.1007/s1247-011-0161-x. https://doi.org/10.1007/s12471-01100161-x.
[30] W. Li, B. Tan, R. Piechocki, Passive radar for opportunistic monitoring in E-health applications, IEEE J. Transl. Eng. Health Med. 6 (2018) 1–10, https://doi.org/10.1109/jtehm.2018.2791609.
[31] A. Wang, D. Nguyen, A.R. Sridhar, S. Gollakota, Using smart speakers to contactlessly monitor heart rhythms, Comm. Biol. 4 (1) (2021), https://doi.org/10.1038/s42003-021-01824-9.
[32] G. Lu, B. Fei, Medical hyperspectral imaging: a review, 2014, J. Biomed. Opt. Society of Photo-Optical Instrumentation Engineers https://www.ncbi.nlm.nih.gov/pmc/articles/PMC3895860/.
[33] E.L. Larsen, L.L. Randeberg, E. Olstad, O.A. Haugen, A. Aksnes, L.O. Svaasand, Hyperspectral imaging of atherosclerotic plaques in vitro, J. Biomed. Opt. 16 (2) (2011) 026011, https://doi.org/10.1117/1.3540657.
[34] A. Hegyi, J. Martini, Hyperspectral imaging with a liquid crystal polarization interferometer, Opt. Express 23 (22) (2015) 28742, https://doi.org/10.1364/oe.23.028742.
[35] ColorIR™ Nir Tunable Filter—Unispectral, UNISPECTRAL, April 22, 2021. https://www.unispectral.com/product/mems-nir-unispectral-filter.
[36] M. Ruzicka, A. Akbari, E. Bruketa, J.F. Kayibanda, C. Baril, S. Hiremath, How accurate are home blood pressure devices in use? A cross-sectional study, PLoS One 11 (6) (2016), https://doi.org/10.1371/journal.pone.0155677.
[37] S.K. Longmore, G.Y. Lui, G. Naik, P.P. Breen, B. Jalaludin, G.D. Gargiulo, A comparison of reflective Photoplethysmography for detection of heart rate, blood oxygen saturation, and respiration rate at various anatomical locations, Sensors 19 (8) (2019) 1874, https://doi.org/10.3390/s19081874.
[38] G. Slapničar, N. Mlakar, M. Luštrek, Blood pressure estimation from photoplethysmogram using a Spectro-temporal deep neural network, Sensors 19 (15) (2019) 3420, https://doi.org/10.3390/s19153420.
[39] J. Wu, M. Dong, S. Santos, C. Rigatto, Y. Liu, F. Lin, Lab-on-a-chip platforms for detection of cardiovascular disease and cancer biomarkers, Sensors 17 (12) (2017) 2934, https://doi.org/10.3390/s17122934.
[40] R. Dhingra, R.S. Vasan, Biomarkers in cardiovascular disease: statistical assessment and section on key novel heart failure biomarkers, Trends Cardiovasc. Med. 27 (2) (2017) 123–133, https://doi.org/10.1016/j.tcm.2016.07.005.
[41] ScienceDirect, Volatile Biomarkers, ScienceDirect, 2013 (Accessed 11 February 2022) https://www.sciencedirect.com/book/9780444626134/volatile-biomarkers.
[42] K. Pariaszewska, M. Młyńczak, W. Niewiadomski, G. Cybulski, Digital stethoscope syste the feasibility of cardiac auscultation, in: Photonics Applications in Astronomy, Communications, Industry, and High-Energy Physics Experiments 2013, 2013, https://doi.org/10.1117/12.2032161.
[43] J.E. Hall, A.C. Guyton, M.E. Hall, Guyton and Hall Textbook of Medical Physiology, Elsevier, Philadelphia, PA, 2021.
[44] J. Allen, Photoplethysmography and its application in clinical physiological measurement, Physiol. Meas. 28 (3) (2007), https://doi.org/10.1088/0967-3334/28/3/r01.
[45] M.J. Daemen, The heart and the brain: an intimate and underestimated relation, Neth. Hear. J. 21 (2) (2013) 53–54, https://doi.org/10.1007/s12471-012-0371-x.

[46] C. Chan, W. Hosanee, Z. Kyriacou, J. Allen, et al., Multi-site photoplethysmography technology for blood pressure assessment: challenges and recommendations, J. Clin. Med. 8 (11) (2019) 1827, https://doi.org/10.3390/jcm8111827.
[47] S. Mejhoudi, R. Latif, A. Toumanari, A. Elouardi, W. Jenkal, Implementation of the algorithms of ECG signal processing on embedded architectures, in: Proceedings of the Mediterranean Symposium on Smart City Application—SCAMS '17, 2017, https://doi.org/10.1145/3175628.3175632.
[48] M. Elgendi, On the analysis of fingertip photoplethysmogram signals, Curr. Cardiol. Rev. 8 (1) (2012) 14–25, https://doi.org/10.2174/157340312801215782.
[49] B. Bent, B.A. Goldstein, W.A. Kibbe, J.P. Dunn, Investigating sources of inaccuracy in wearable optical heart rate sensors, NPJ Digital Med. 3 (1) (2020), https://doi.org/10.1038/s41746-020-0226-6.
[50] M.R. Montinari, S. Minelli, The first 200 years of cardiac auscultation and future perspectives, J. Multidiscip. Healthc. 12 (2019) 183–189, https://doi.org/10.2147/jmdh.s193904.
[51] I. Bank, H.W. Vliegen, A.V.G. Bruschke, The 200th anniversary of the stethoscope: can this low-tech device survive in the high-tech 21st century? Eur. Heart J. 37 (47) (2016) 3536–3543, https://doi.org/10.1093/eurheartj/ehw034.
[52] S. Leng, R.S. Tan, K.T. Chai, C. Wang, D. Ghista, L. Zhong, The electronic stethoscope, BioMed. Eng. OnLine 14 (1) (2015), https://doi.org/10.1186/s12938-015-0056-y.
[53] S. Swarup, A. Makaryus, Digital stethoscope: technology update, Med. Devices Evid. Res. 11 (2018) 29–36, https://doi.org/10.2147/mder.s135882.
[54] M. AlGhatrif, J. Lindsay, A brief review: history to understand fundamentals of electrocardiography, J. Community Hosp. Intern. Med. Perspect. 2 (1) (2012) 14383, https://doi.org/10.3402/jchimp.v2i1.14383.
[55] L. Kersey, K. Lilly, N. Park, ECG heartbeat classification: an exploratory study, in: Proceedings of the Australasian Joint Conference on Artificial Intelligence—Workshops, 2018, https://doi.org/10.1145/3314487.3314491.
[56] M.A. Serhani, H.T. El Kassabi, H. Ismail, A.N. Navaz, ECG monitoring systems: review, architecture, processes, and key challenges, Sensors 20 (6) (2020) 1796, https://doi.org/10.3390/s20061796.

Further reading

F.S. Cikach, R.A. Dweik, Cardiovascular biomarkers in exhaled breath, Prog. Cardiovasc. Dis. 55 (1) (2012) 34–43, https://doi.org/10.1016/j.pcad.2012.05.005.
K. Bartels, R.H. Thiele, Advances in photoplethysmography: beyond arterial oxygen saturation, Can. J. Anesth. 62 (12) (2015) 1313–1328, https://doi.org/10.1007/s12630-015-0458-0.
A.A. Kamshilin, N.B. Margaryants, Origin of photoplethysmographic waveform at green light, Phys. Procedia 86 (2017) 72–80, https://doi.org/10.1016/j.phpro.2017.01.024. International Conference on Photonics of Nano- and Bio-Structures.
A.E. Aubert, B. Verheyden, Neurocardiology: a bridge between the brain and the heart, Biofeedback 36 (1) (2008) 15–17. https://www.aapb.org/files/publications/biofeedback/2008/biof_neurocardiology.pdf.
Non Communicable Diseases, World Health Organization, https://www.who.int/en/news-room/fact-sheets/detail/noncommunicable-diseases.

M. Campbell, A. Sultan, L. Pillarisetty, Physiology, Korotkoff Sound, StatPearls. U.S. National Library of Medicine, July 12, 2021. https://www.ncbi.nlm.nih.gov/books/NBK539778/.

Sphygmomanometers. 3. How hard is it measure? https://ec.europa.eu/health/scientific_committees/opinions_layman/sphygmomanometers/en/l-3/3.htm.

Review—Non-Invasive Monitoring of …—Institute of Physics, 2020 (Accessed 11 February 2022) https://iopscience.iop.org/article/10.1149/1945-7111/ab67a6/pdf.

CHAPTER 3
Sensor signal analytics

1. Introduction

In Chapter 2, we introduced multiple sensors of differing form factors ranging from medical grade sensors to the sensors embedded in smartphones and wearables. These sensors record different physical signals such as electrical parameters (voltage, current), sound (pressure), and motion (accelerometer), all of which have a direct (or indirect) correlation to certain cardiac activities and/or markers. Signals from these sensors need further analysis in order to track multiple cardiac activities. In this chapter, we focus on the processing steps that are used to analyze the sensor signals. In Fig. 1, we present a generic end-to-end signal processing flow diagram. Please note that this chapter entails a generic description of the analytic tools for cardiovascular analysis for sensor signals along with a process flow for their application. Given the nature of the data we cover in this book, the processing steps are limited to one dimensional (1D) digital health data analysis. The detailed processing steps for a specific combination of sensor(s) and associated medical marker(s) and/or condition(s) are detailed in each chapter of Section 2. This chapter provides a basic approach that is the prerequisite of the information found in Section 2.

2. Preprocessing

As the name suggests, preprocessing is a step before the core of the processing begins. Pick any sensor, and its data will be corrupted with noise and artifacts. Let us take the example of a heart rate monitor that is commonly available in smartwatches. Typically, a photoplethysmography (PPG) sensor is attached to the inner wrist. Being a reflective sensor, the recording will be solely dependent on the infrared (IR) and visible red lights that are reflected from the wrist. Given that these watches are meant to be worn all day, every day, all the body movements will find their way into the recorded PPG signal. The data needs to be cleaned and prepared, and there is no one-size-fits-all solution, even for a single sensor. The steps and degree

New Frontiers of Cardiovascular Screening using Unobtrusive Sensors, AI, and IoT
https://doi.org/10.1016/B978-0-12-824499-9.00003-9

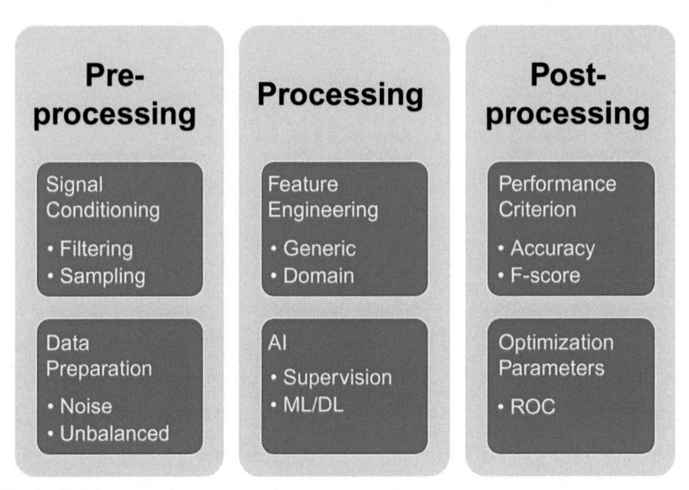

Fig. 1 Block diagram.

(aggressiveness) of the preprocessing will vary depending on the desired outcome of the flow. If we want to calculate the heart rate and heart rate variability, we may use a more aggressive technique of noise/artifact removal, even if that entails smoothing of the PPG signal, as we do not need to retain the finer details of the PPG morphology. However, if the requirement is to predict a complicated cardiac marker and/or disease such as coronary artery disease (CAD) (see Chapter 5), we may need to track the notch location of the PPG. In this case, the preprocessing cannot be too aggressive as it may also smooth out the disease features. The key to designing an effective preprocessing pipeline is to understand the desired signal characteristics and implement a matched filter to the desired signal. We describe these preprocessing steps in detail in subsequent sections.

2.1 Signal conditioning

Signal conditioning refers to the processing of an analog signal before it is digitized and passed on to the next stage. Most analog signals need to be prepared prior to digitization. For an example, an amplifier may be needed to enhance small voltage recordings. The amplification factor will depend on the intended output range of the signal. Below are the common steps involved in signal conditioning.

2.1.1 Hardware filtering

Hardware filters can be categorized in many ways. One basic categorization is electronic vs nonelectronic filters. Nonelectronic filters can be a mechanical design which intrinsically enhances the signal-to-noise ratio (SNR). A common example of such filters would be the design of a stethoscope (see Chapter 2). Nonelectronic filters in the current scope of discussion are mostly passive, i.e., they do not require an external power source. Electronic filters can be further categorized into passive and active filters. Active filters are those which require external power. An active filter will have active components such as an amplifier. On the other hand, passive filters consist of components such as resistors, capacitors, inductors, and transformers.

With the advent of technology, processing resources have become quite frugal in past few decades. This has led to many embedded off-the-shelf devices that have dedicated onboard hardware and software filters. For example, do you know how many sensors are there in common smartphone of 2020? More than 20. Each comes with dedicated set of filters which are applied before the signal is available for processing. Here we mention some of the common filtering steps.

2.1.2 Amplification

Let us examine a fingertip pulse oximeter. This records two PPGs in the form of voltage signals (for IR and visible red wavelengths). The voltage signals are proportional to the amount of oxygenated and deoxygenated blood flowing through the blood vessels. Those voltage signals are processed in turn to arrive at the final parameter, blood oxygen saturation (SpO_2). Now let us look at some of the output requirements: (a) SpO_2 is a normalized percentage (theoretical range is 0%–100%); (b) if SpO_2 falls below 88%, the user needs to be put on external oxygen support. These two requirements govern the SpO_2 sensitivity requirement and one can reverse-calculate to arrive at the amplification factor needed for the voltage signals.

2.1.3 Attenuation

Attenuation is the opposite of amplification. Sometimes an electrical signal, typically voltage, needs to be reduced to in order to ensure that the processed signal is within the analog to digital converter (ADC) range.

2.2 Noise handling

It is a common notion in conventional signal processing to treat the noise before applying the rest of the processing steps. However, it is more practical to treat the noise source(s) as different sensors and with different sources. It is also important to treat the noise as early as possible in the analytics pipeline. As shown in Fig. 2, some of these noises cannot be changed, which leaves us

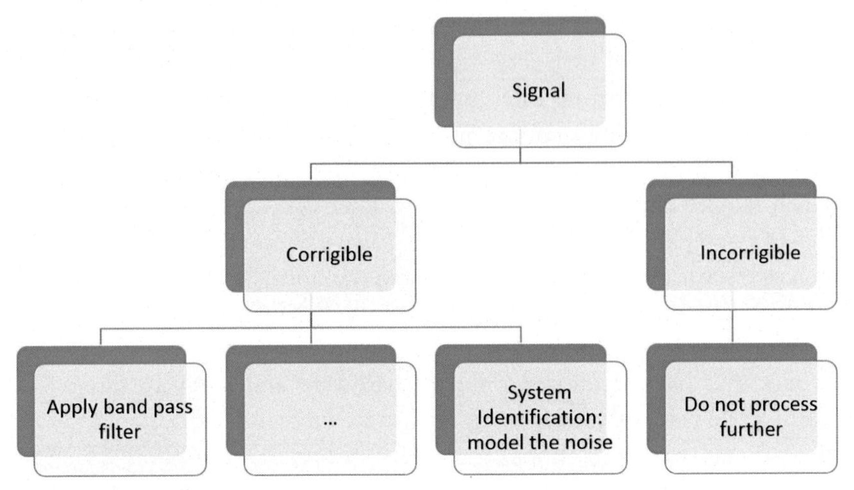

Fig. 2 Noise handling.

no choice but to mark them as unusable. This is important as processing a low signal-to-noise ratio (SNR) data will yield erratic inconsistent outputs with potentially dire repercussions. Imagine the difference in someone's heart rate if it is reported as either 300 or 30 due to the movement noise induced in his/her heart rate measurement device (e.g., a corrupted PPG in a smartwatch). On the other hand, there will be noisy signals which have a low SNR, but there are a few techniques to correct them. Noise reduction techniques can be broadly categorized into noise removal and signal reconstruction.

2.2.1 Noise sources

Noise sources are of a different nature. This can result from the hardware, software, or background noises. Here we mention some of the common noise sources.

- Hardware: Leakage current flows in an unwanted conductive path. It cannot be made to be absolute zero. Currently there are safety standards in place which specifies the maximum permissible leakage current for different grades of devices. However, that step is mostly meant for the safety of the user and the recorded signal is still distorted.
- Ambience: A stethoscope or any other sound-capturing device will capture anything permitted by the microphone range (generally 20 Hz to 20 kHz). Hence, a stethoscope used in a noisy environment (busy rural hospital or self-monitoring) will capture unwanted noises which in turn will distort the heart sounds (S1 and S2, detailed in Chapter 5).

- Software: Any operation on the signal introduces noise. The trick is to ensure that the operation cleans up more than it corrupts and in the process, tip the scale in favor of higher SNR. For example, each of these 1D sensors often needs to be filtered in a predetermined frequency range.
- Usage: Usage of sensors leads to signal corruption. For example, when wearable devices and smartphones are used when the user is moving, more often than not, the movement is captured in the sensor signal. However, it is important to note that not all the signals become corrupted in the same manner or degree. For example, let us assume a person is running with a smartphone with a PPG sensor and he/she also has an electrocardiogram (ECG) band strapped to the chest. The PPG signal is greatly susceptible to noise corruption compared to the ECG signal.

2.2.2 Noise removal

A vast array of digital filters exist to remove noise from the time-series signals. As shown in Fig. 3, depending on the frequency range of signal frequency and noise frequency, one may choose the appropriate filters. However, the following factors need to be considered while designing a filter:

1. Ripple effect: A Butterworth filter is often chosen as it produces no amplitude ripple in passband or stopband. The downside of a Butterworth filter is that it provides very wide transition region.
2. Transition: A filter with a steeper transition is preferred, provided the other parameters are unaffected or relatively less distorted. Amongst the popular filters, Chebyshev filters provide the opportunity to control

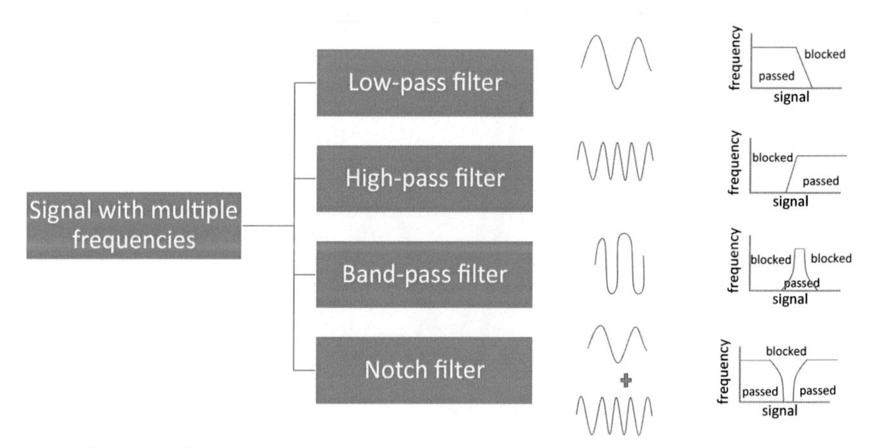

Fig. 3 Different filters.

ripple (for example, we might consider a 3 dB ripple is acceptable). Furthermore, the Chebyshev filter design provides a handle to choose the ripple in the pass band (Chebyshev Type I) or the stop band (Chebyshev Type II). The advantage of Chebyshev over Butterworth is that it provides steeper transition region roll-off.

3. Filter order: With increasing filter order, the frequency magnitude response performance improves. Hence, if someone aims to design steep transition, and lesser transition band, the filter order should be increased. However, a higher filter order also introduces instability (in an infinite infrared [IIR] filter) and processing time (in a finite infrared [FIR] filter).
4. IIR or FIR: FIR filters can be designed to produce a linear phase whereas IIR produces a nonlinear phase. FIR filters have a guaranteed stability which does not exist in IIR filters. The fixed point calculations introduce quantization errors which have less impact in FIR filters. However, this comes at a cost of higher computational cost and memory requirements. IIR filters are typically low latency when compared to FIR filters, and are hence preferred for real-time operations. Fig. 4 shows the responses of both of the filters.

2.2.3 Signal reconstruction

Incorrigible noise needs to be removed. However, often while processing time-series signal, one may finding a coarse marker by reconstructing a noisy signal. This is particularly useful in the scenario where the nature of the noise is dynamic. One popular choice of signal reconstruction in case of variable noise is an adaptive filter. As shown in Fig. 5, an adaptive filter is a type of digital filter which takes two inputs: Primary input $d(n)$ and a reference

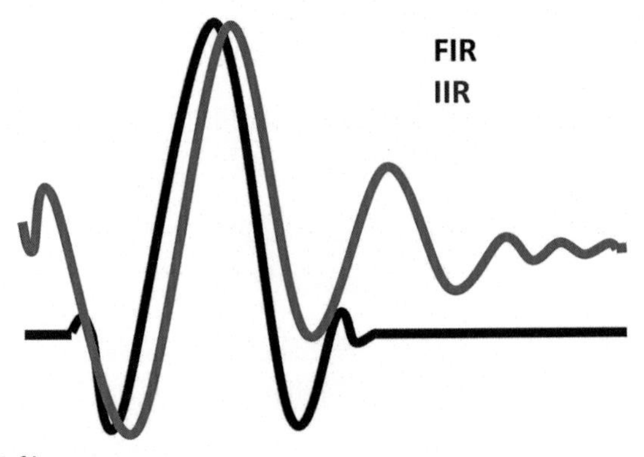

Fig. 4 FIR vs IIR filter response.

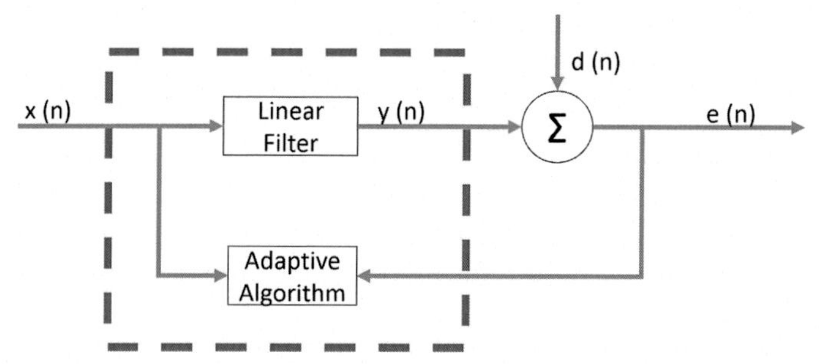

Fig. 5 Adaptive filter.

input $x(n)$. The adaptive algorithm tries to set the filter parameters so that the reference signal postmodification comes close to the residue signal $e(n)$.

For example, the PPG signal embedded in a smartwatch is affected by fairly destructive motion artifacts as compared to the PPG signal recorded by an oximeter. However, one can apply advanced signal processing techniques such as recursive-least-square (RLS) filtering and singular spectrum analysis (SSA) in order to reconstruct parts of it [1]. This provides more accurate detection of beat-to-beat cardiac intervals leading to better tracking of heart rate variability. This technique can help to monitor 24×7 heart rhythms of a person, resulting in the detection of brief atrial fibrillation (AF) episodes which are usually not detectable without employing invasive methods. However, the reconstructed PPG will typically be oversmoothened, thereby losing its finer morphological features such as the position of the dichroitic notch. Hence, this PPG cannot be used to extract complex discriminative features used in CAD classification.

3. Decision making using AI

Artificial Intelligence (AI) has experienced resurgence in the 21st century following the concurrent advances in computing power, the attainability of large volume of data, and theoretical understanding. AI has become an essential part of our everyday life, helping to solve many challenging problems in computer science, software engineering, and operations research. Applications of AI can be extensively found in the field of biomedical engineering in the form of designing of decision support systems for prediction of diseases from large amounts of data. Medical data are often huge in volume and are typically recorded 24×7 for the prediction of intermittent diseases such as an AF episode or presence of an arrhythmia from an ECG recording of prolonged duration. It is practically impossible for the doctors to analyze such a large volume of data. Here, the AI-based systems play an important role in analyzing such data in a short period of time by a computer to produce meaningful information. The application of AI that enables a computer to automatically learn from data without human intervention is called machine learning (ML). Machine learning focuses on the development of computer programs to analyze input data provided as input to perform a desired task.

Machine learning techniques can be broadly categorized into supervised and unsupervised approaches.

3.1 Supervised machine learning

As the name suggests, the computer learning is supervised to ensure the information is being learnt correctly. This means that when the computer is fed with a dataset as input to learn, it must be supplied with the correct answer-sheet that contains the possible output of every instance of the input dataset. In ML, the input data to the computer is known as the features and the answer-sheet is termed the output labels/targets. The input dataset together with the output labels is called the training dataset. All the input data corresponding to one output label is termed a class. Every superseded learning technique has two fundamental phases, training and testing. In supervised learning, an algorithm is employed to learn the mapping function from the input data (x) to the output labels (y); that is $y = f(x)$. The objective of this learning is to approximate the mapping function (f). During training, the computer is provided with a training dataset and the training process returns a set of numerical values that defines a mathematical relationship between the input features and the targets, which is called the training model. This training model is the output of the training phase. In the test phase, with the help of this training model, the computer can successfully predict the outcome of unlabeled and unseen input features, which are called the test data.

Supervised learning algorithms are developed to perform two fundamental tasks, classification and regression.

A classification job aims to identify a set of input features to one of the two or more labels used to train the system. The output label values are discrete and categorical. An algorithm that can identify the presence of a cat image from an unlabeled snap is a typical example of a classification problem. On the other hand, in the case of a regression job, the output labels are continuous numerical values. An algorithm that can estimate a house price based on the floor area, and number of rooms, in each locality, based on the training data, is a typical example of a regression problem.

3.2 Unsupervised machine learning

In unsupervised ML only the input features are applied to the system and no labels are provided. As expected, no definite training is possible under an unsupervised approach. The algorithms are intended to infer a function to describe a hidden structure from these unlabeled data. Although the machine cannot figure out the correct output, it analyzes the data and can draw inferences from the features to describe the hidden structure. Applications of supervised ML techniques are popular in cardiology for various tasks such as the classification of control diseased persons from biomedical signals such as ECG or the prediction of cardiovascular risk of a person based on his past and family disease history and physiological vitals such as heart rate, blood pressure, blood cholesterol, and blood glucose.

Fig. 6 shows the generic block diagram comprising various fundamental steps of a classification system that can be applied for various cardiac applications such as the clarification of normal cardiac patients based on biomedical signal analysis. The first two blocks, data acquisition, preprocessing and various noise cleaning techniques have already been discussed in detail. In this section, we will discuss the remaining steps for performing a classification task.

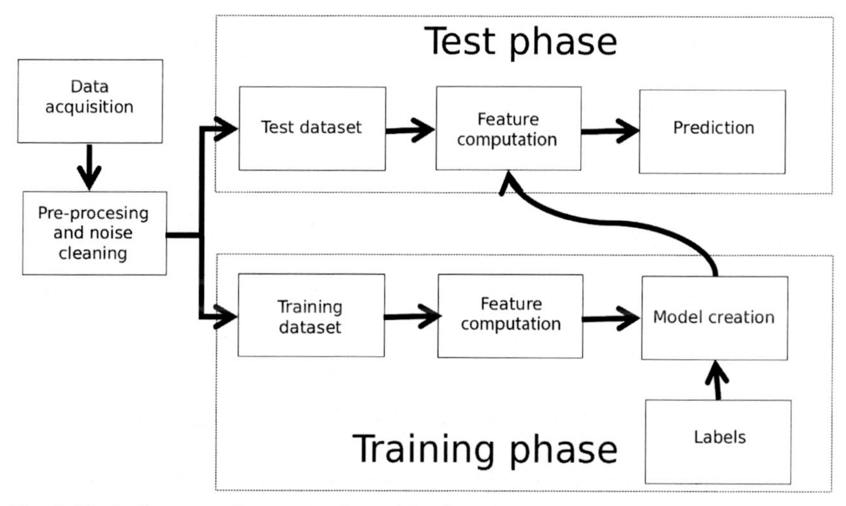

Fig. 6 Block diagram of supervised machine learning system.

3.3 Splitting of training and test data

This is an important step in any classification task. First, it is important to note that both the training and the test sets should be diverse enough to individually represent the statistical properties of the entire dataset. Failing to do so will either cause an imperfect training model or an incomplete test process. Second, the training and the test data should not be mutually exclusive. In other words, the underlying statistics of the training and the test data should be similar. It is also recommended to maintain the distribution of different classes in the training and the test sets identical to the original dataset. The goal of an ideal training is to create a model that generalizes well to the test data. However, in a practical scenario, the training data is often contaminated by noise. Sometimes there are scenarios when training becomes so exhaustive that the model learns the noise in the training data, negatively impacting the classification performance on the test data. This is called an overfit and should be avoided. Likewise, if training does not ensure the model learns the underlying pattern of the training data, the generated model is inevitable to fail to classify the new data. This condition is called an underfit [2].

In order to mitigate the chance of overfit and underfit, a popular approach is to split the entire dataset into three parts, namely the training, test, and the validation set. The validation set is used as an intermediate test set to evaluate the performance of the model after each training session and the learning process is adjusted accordingly. Once the model yields satisfactory performance on the validation set, it is released as the optimum model which is evaluated on the test set. The splitting of the training-validation-test sets can be done by random selection from the entire dataset. Typically, 60%–70% of the entire dataset can be taken for training, 10%–20% for internal validation, and the rest for test purposes.

3.4 Feature engineering

Feature engineering is undoubtedly the most difficult and time-consuming part of any ML task. Researchers typically invest most of their time in identifying the proper feature set that results in the optimum classification performance. A classification algorithm takes a set of images or time-series data as input. In our case, the input signals are typically digitalized biomedical signals such as a set of ECG time series. The digital time-series data acquired by the sensors are an array of numbers indicating a voltage measurement at a given timestamp. The time series does not have any statistical or physiological significance in general. Hence, applying the time-series data directly to a ML algorithm does not yield any meaningful analysis. Hence, we need to create a meaningful representation of the time series that contains the underlying properties of the time series, which is called the feature. A feature can be a single numerical value or a multiple of many values. In the latter case the representation is called the feature vector. The task of feature engineering is to identify a feature vector that is significantly different for different classes in the training dataset, so that the learning algorithm can determine the boundary for different classes. This is called the relevant or the discriminating feature vector. The success of a ML task strongly depends upon identifying the relevant feature vector for the task. For example, the morphology ECG signals recorded from a normal noncardiac subject and an arrhythmia patient may often look similar. In the following stage, we measure the mean heart rate value from every ECG signal from the successive R peak to R peak interval distances.

Arrhythmia patients typically exhibit an abnormal heart rate which is above 100 bpm or below 60 bpm owing to underlying pathological conditions. On the other hand, the heart rate of healthy people typically ranges between 60 and 80 bpm. However, arrhythmia patients often experience random fluctuation in successive cardiac intervals. Hence, the standard deviation of the R-R interval distances typically becomes more for the arrhythmia patients. Hence, we can easily classify the normal subjects and arrhythmia patients from their ECG signals based on a 2D feature vector comprising the mean heart rate and the standard deviation of the RR interval distances in the recording. Biomedical signals such as ECG, phonocardiogram (PCG), and PPG are periodic but nonstationary time-series signals. There are certain parameters that are considered as meaningful features as they have been successfully implemented in a number of ML applications. The features can be broadly categorized as time domain features. They are extracted from the time series and typically capture the morphological properties of the signals. A few such features are listed below [3].

(1) Mean periodicity: Most of the biomedical signals are periodic in nature and the mean peak-to-peak interval distances (the R peaks of an ECG, the S1 or the S2 peaks of a PCG, etc.) indicate the mean heart rate.

(2) Heart rate variability: The instantaneous heart rate changes over time. Even for a young healthy adult with no prior cardiac history, the instantaneous heart rate slightly fluctuates over time. The rate of fluctuation typically increases under physical stress or mental anxiety. This is called the heart rate variability (HRV). The extent of HRV typically becomes higher than normal due to various types of arrhythmia and becomes lower than normal due to arterial blockages. There are a set of parameters for mathematically quantifying the HRV [4,5]:

- **(a)** Standard deviation of NN intervals (SDNN): This is the standard deviation of the successive peak to peak intervals (clinically termed the NN intervals) typically measured in a period of 24h. There is another parameter called standard deviation of the average NN intervals that is typically measured for 24h.
- **(b)** Root mean square of successive distances (RMSSD): This is measured by the square root of the mean of the square value of the distances between the successive NN interval distances.
- **(c)** Standard deviation of successive distances (SDSD): Measures the standard deviation of the distances between the successive NN interval distances.
- **(d)** NN50: The number of pairs of successive NN intervals that differ by more than 50ms.
- **(e)** pNN50: The value of NN50 normalized by the total number of NN intervals in the recording.
- **(f)** NN20: the number of pairs of successive NN intervals that differ by more than 20ms.
- **(g)** pNN20: The value of NN20 normalized by the total number of NN intervals in the recording.

(3) Other morphological features, including PR interval, QT interval, QRS interval distances measured from an ECG cycle; rising slope and falling slope of a PPG pulse; S1-S2 interval and S2-S1 interval distances measured from a PCG cycle [6] are considered to be important features in various applications.

3.4.1 Dimensionality reduction using principal component analysis

So far, we have discussed the basic flow in a ML system. Acquiring and cleaning of the raw data is the first challenge. Once done, the next big step is extraction of relevant features. The more features that are extracted, increases the likelihood that the classifier is able to classify in a better way. However, increasing the feature dimension may not always make sense. For example, suppose you want to classify arrhythmia based on ECG. You have the domain knowledge that a persistent heart rate value above 100Hz can be classified as arrhythmia. Now, you decide to extract a 2-dimensional feature vector, R-R interval distances and P-P interval distances from the ECG. However, these two features contain almost similar information. Applying both features to the classifier will unnecessarily increase the lead of the classifier. Let's consider a more practical scenario; you have extracted 100 features to train a supervised machine classifier, but 30 of them carry similar information. In that case it will take more time to train your classifier and the model size would be large. Hence, we use a procedure called dimensionality reduction to effectively eliminate the redundant features before applying the data to a classifier.

Principal component analysis (PCA) is a simple yet effective dimensionality reduction technique. The objective of PCA is to project the data into a smaller space in order to reducing the dimensionality of our feature space [7,8]. It tries to extract the important information from the data and to express this information as a set of summary indices called principal components. The principal components have nothing to do with the original features. The newly formed transformed feature set or the principal components will have the maximum variance. The second principal component will have the second-highest variance and so on. Let's get back to our original example of two features. PCA tries to find the best fit line for these data points which minimizes the distance between them and their projections on the best fit line. This is done using the singular value decomposition (SVD) technique. As the principal components are orthogonal to one another, they are statistically linearly independent of one another.

3.4.2 Independent component analysis

Independent component analysis (ICA) is a popular technique to separate independent sources from a mixed signal. Unlike principal component analysis, which focuses on maximizing the variance of the data points, independent component analysis focuses on independent components in a mixed signal, e.g., separating two voices simultaneously talking at a party. ICA assumes that hidden independent components in a mixed signal are statistically independent and are non-Gaussian. ICA is related to principal component analysis but is a much more powerful technique. However, ICA is capable of finding the underlying factors or sources when these classic methods fail completely. The detailed mathematical explanation for ICA is provided in [9,10].

3.5 Popular machine learning algorithms

So far, we have discussed feature extraction and dimensionality reduction. Features are the discrete numeric values that represent the salient properties of a data in a reduced space. The features are fed to a block that classifies them to a target label. A supervised learning approach typically involves a training phase. Here, it takes the features and their corresponding labels as input and tries to relate them. The output of the training is a set of numbers called the training model which can be used to predict a new instance of unseen feature values, termed the test data. In this section, we will briefly discuss popular ML algorithms.

3.5.1 Logistic regression

Logistic regression is a predictive model popularly used for binary and multi-label classification. It tries to predict the probability of a category-dependent variable which are the target labels based on a set of independent feature values. The target value is a binary variable 1 or 0 for a given feature set X and target value Y, where logistic regression predicts the probability $P(Y=1)$ as a function of X. A detailed description and the underlying mathematical analysis can be found in [11,12]. Multiclass classifier logistic regression takes one type of target label as 1 and merges the rest as 0 and repeats for all classes. This is called the one vs rest approach.

3.5.2 Support vector machine

Support vector machine (SVM) is probably the most popular algorithm used in supervised ML. For two classes, SVM tries to create a hyperplane that maximally separates the data-points corresponding to the two class labels [13,14]. The data-points corresponding to the classes which are the closest to the hyperplanes are called the support vectors. Only the support vectors decide the position of the hyperplane. SVM works very well in a high dimension features space. It is effective even if the feature dimension is greater than the number of samples in the training data. Because it uses a subset of training points in the decision function to create the hyperplane (called support vectors), it is a memory-efficient algorithm.

3.5.3 Decision tree

A decision tree is a nonparametric supervised learning method. The objective is to create a model that predicts the value of a target variable by learning simple decision rules inferred from the features [15]. A decision tree is a flowchart-like tree structure where an internal node represents a feature, a branch in the tree represents a rule, and the leaf nodes represent the outcome of that rule. The topmost node in a decision tree is called the root node. It learns to partition on the basis of the feature value. Then, it partitions the tree in a recursive manner to reach the leaf node. This flowchart-like structure helps in final prediction. One advantage of decision tree is that it works well on both numerical features and also on categorical features.

A decision tree can overfit. Instead of a creating a single decision tree, multiple decision trees can be formed by partitioning the whole training data. Combining their outputs using a majority voting can avoid an overfit. This is called the random forests algorithm [16].

3.6 Deep learning in biomedical engineering

So far, we have discussed the traditional approaches of classifying disease events using AI and ML. To summarize, in order to classify the presence of a disease from an input signal, we first need to identify relevant features that are substantially distinguishable between the classes we want to recognize. The performance of such an approach widely depends upon the skill of identifying the set of discriminating parameters that can effectively separate the two classes. Identifying such features is arguably one of the most difficult and time-consuming tasks in a traditional ML approach. It often requires medical knowledge and data visualization to identify the features. Hence, researchers have been considering alternative approaches to remove the manual feature detection required to train a classifier.

Deep learning is a reasonable solution to the problem. Deep learning is often considered a new established norm of modern data science. The fundamental principle in deep learning is the user does not have to worry about any manual feature computations. The deep learning itself maps the input data into a higher dimension data space for efficient spatiotemporal analysis. In deep learning, the feature mapping is often performed using a set of filters, commonly known as kernels. The filters typically perform a mathematical operation which is called a convolution.

The convolution is the basic operation of deep learning where a rectangular filter window is placed on top of the input data, multiplied with the input, the result is stored, the filter is shifted and multiplied, and the result is appended. A spatiotemporal feature mapping is possible from the input, thanks to the convolution operation. This type of deep learning technique is called the convolution neural network (CNN) and it is by far the most popular deep learning technique.

CNNs are popularly used for image classification and image recognition where 2D spatial feature extraction is required. However, they are equally popular in 1D time series analysis, e.g., speech synthesis, speaker identification, or processing of biomedical signals.

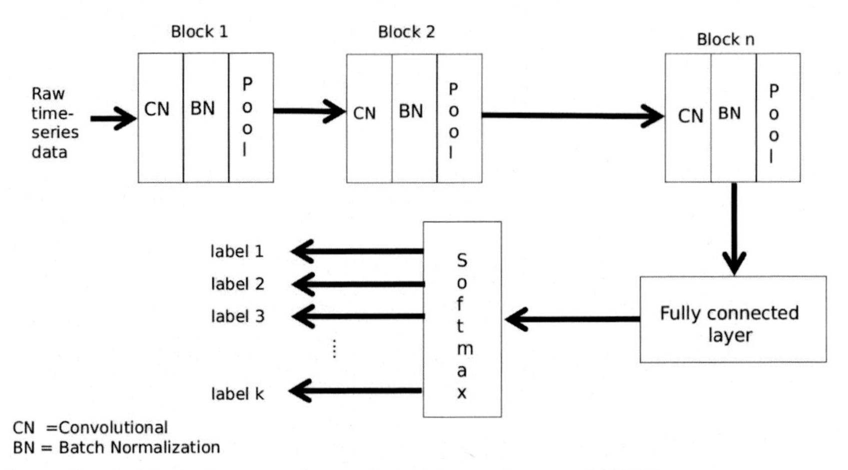

Fig. 7 Simple block diagram of convolutional neural network (CNN).

The simplest block diagram of a CNN architecture for classification of 1D signals is shown in Fig. 7. This type of architecture can be used for classification of normal subjects and cardiovascular patients, considering the biomedical data recorded from them as input. Popular examples of such applications can be classifying AF episodes from ECG time-series data or identifying cardiac murmurs in digitally recorded heart sound data.

As the input data is 1D, the filters for performing the convolution operation are also done in the 1D domain. The architecture has a set of blocks, each of them comprising three modules: convolution, batch normalization, and pooling. We have already discussed convolution. The output of the convolutional layer is a higher dimensional feature map. This is passed through an activation function which determines which outputs will go to the next layer. Nonlinear activation is typically performed in deep learning. Rectified linear unit (ReLU) has been a popular choice of activation function. It is a piecewise linear function where the input is directly passed as output, in the case that it is positive, and the output is zero in the case that the input is negative.

The batch normalization performs a scaling operation on the convolution output. This procedure is important to optimize the training process. The pooling layer takes a representative output of the normalized data based on small windows. The maximum or the average value of a window is typically considered as the representative value of a window. The pooling layer is responsible for reducing the high dimensional feature map created by the convolution filter to a low-dimensional space to reduce the subsequent computation load. The output of one pooling layer is applied to the input of the next block and the same set of operations are performed. The output of each convolutional block is flattened in the fully connected layer. There can be one or several such fully connected layers. The final block in the CNN is the softmax layer which assigns the probability for each target class that determines the final classification labels [17,18].

The aim of deep learning architecture is to learn the optimum weights corresponding to different layers in an iterative manner, which is known as training. During this phase, the deep learning engine is provided with training data and the corresponding labels which are called ground truth values. In order to learn the optimum set of weights, a deep network first assigns random weights to all its layers and checks against its training data as to how much the validation is from its ground truth or from its predicted value. As the initial weights are randomly chosen, it is expected that the predicted labels will hardly match the ground truth values. In the subsequent stages, the weights are modified, and the corresponding impacts are measured. This is done in an iterative manner. When the performance is optimized, the training is considered complete.

The learning is done through a process called backpropagation algorithms [19]. The backpropagation aims to minimize the loss function, which is a quantitative measurement of how closely the predicted outcomes match the actual values. For regression problems, the loss function can be the mean squared error between the predicted and the actual values. In classification problems, a new type of loss function is used, which is called the cross-entropy loss. Cross-entropy loss measures the performance of a classification model whose output is a probability value between 0 and 1. Cross-entropy loss increases as the predicted probability diverges from the actual label. Hence, minimizing the cross-entropy loss during training ensures that the weights are tuned so that the predicted labels are likely to be similar to the actual labels. The way the weights are changed in iteration is determined by the optimization algorithm. The simplest form of the optimization algorithm is called the gradient descent. This is an iterative first-order optimization algorithm used to find a local minimum/maximum of a given function. There is another parameter called the learning rate that determines the rate at which the weights are minimized at every iteration.

A major concern of the deep learning architectures is that they are prone to overfit because neural networks are large in size, and they can have millions of trainable parameters. A popular choice to avoid overfit is to randomly disconnect a few neuron connections in the network. This process is called dropout [20]. This simple process has been a proven remedy of overfitting that works on any new dataset. Applying the rate of dropout is an important area in deep learning, typically done on a trial and error basis to optimize the performance.

3.6.1 *Neural network activation*

Neural networks behave similar to the human brain. A neural network contains hundreds or thousands of neurons. However, not all of them remain active simultaneously. Only a fraction are active at any one time. A neuron typically has multiple inputs. The weighted sum of the inputs is applied to function which determines whether the input to the neuron will pass to the next stage of the neuron. This function is called the network activation function. Various activation functions can be used depending upon the configuration of the network, as discussed below.

Binary step activation function

As shown in Fig. 8, this is the simplest form of binary activation function that returns 1 if the input is positive and returns zero if the input is negative.

Linear activation function

The input is passed as it is as the network output. The function looks as shown in Fig. 9.

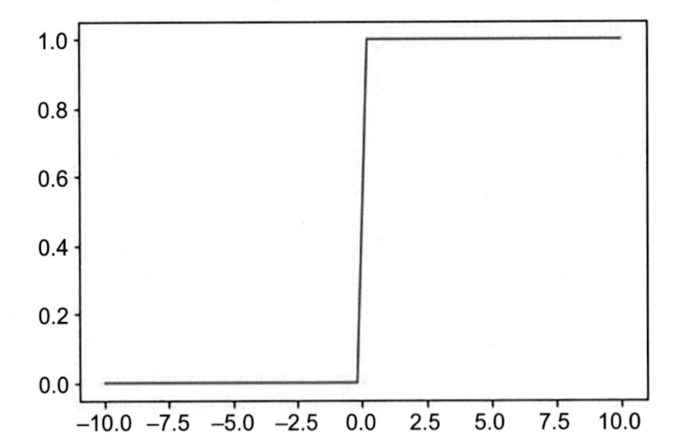

Fig. 8 Binary step activation function.

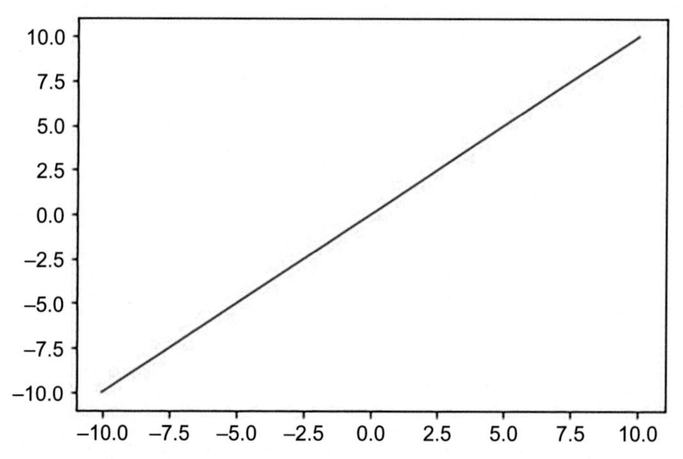

Fig. 9 Linear activation function.

Sigmoid activation function

This is a nonlinear activation popularly used in neural networks. The mathematical expression is:

$$sigmoid(x) = \frac{1}{1 + e^{-x}}$$

The plot is shown in Fig. 10. This function returns a value between 0 and 1. The functions output 0.5 when the input is 0. The output value goes close to 1 as the input increases from 0 and the output goes close to 0 as the input decreases from 0. One disadvantage of sigmoid function is the network

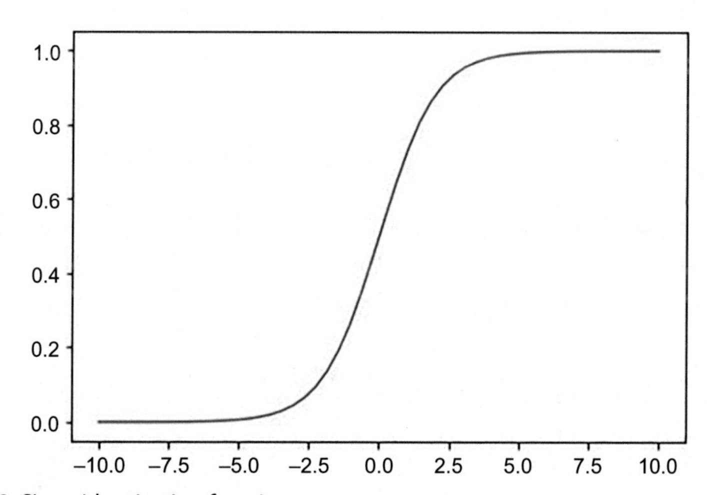

Fig. 10 Sigmoid activation function.

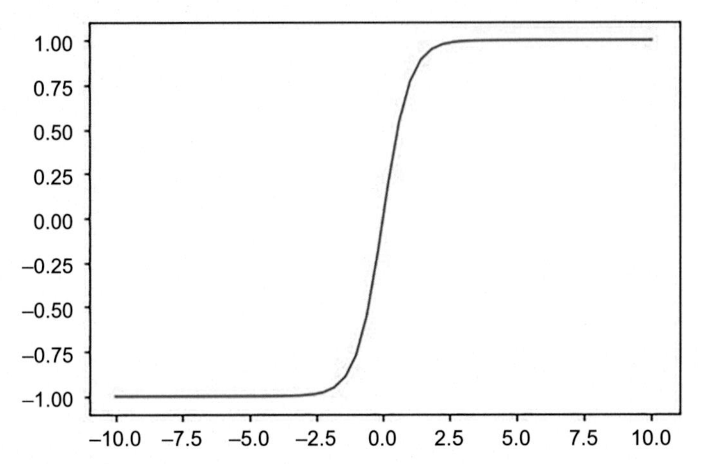

Fig. 11 Tanh activation function.

does not learn quickly near the boundaries. This is because gradient is almost zero near the boundaries.

Hyperbolic tangent activation function (tanh)

This is another nonlinear function similar to sigmoid. However, the output value is between -1 and $+1$. The corresponding plot is in Fig. 11.

$$\tanh(x) = \frac{1 - e^{-x}}{1 + e^{-x}}$$

Rectified linear unit activation function

ReLU is the commonly used activation function in deep learning networks. ReLU can be used to train very deep neural networks. It requires less computational load than the other nonlinear activation functions. It returns 0 if the input is less than 0 or passes the input value as output if the input is greater than 0. In Fig. 12, we have plotted the ReLU activation function.

Softmax function

Softmax returns numeric output of the last linear layer of a multiclass classification neural network into probabilities. It is typically used in the final layer in a multiclass neural network classifier to calculate the assigned probability of different class labels of the input.

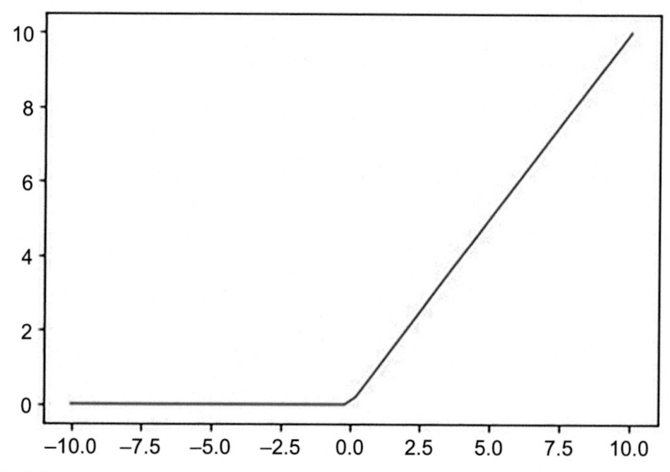

Fig. 12 ReLU activation function.

3.6.2 Network hyperparameters

As discussed in the previous section, neural networks are large, complex in nature, and contain multiple repetitive layers. Designing the optimum neural network to perform a certain task is the most difficult job. However, there are no defined steps to design the best performing network for all types of problem. In fact, no such network exists. It is the responsibility of the programmer to design the best networks that optimally perform the job. As mentioned, a neural network contains many parameters, and each of them is directly responsible to determine the performance of the network. These are called hyperparameters. In order to get the best performance out of a neural network, one must carefully tune them. Some popular hyperparameters are: The number of convolutional layers in the network, kernel dimension, number of filters used in each convolutional layer, rate of dropout, and learning rate. The hyperparameters are set during training and the training model is evaluated on an independent dataset and adjusted accordingly.

There are two popular approaches of tuning the hyperparameters. In a grid search approach, a search space is defined as a grid of hyperparameter values to evaluate every position in the grid. This is an exhaustive searching that checks every possible combination of hyperparameters. On the other hand, in a random search, a search space is defined as a bounded domain of hyperparameter values and random sample points are selected in that domain from different hyperparameters to calculate the optimum combination. This is a relatively faster process.

3.6.3 Recurrent neural networks

So far, we have discussed traditional neural networks such as CNN which are mostly used for spatial feature extraction. They assume that all the inputs and outputs are independent of each other. But in cases such as time-series analysis, for example, when it is required to predict the next word of a sentence, the previous words and the context are required and hence there is a need to remember the previous words (termed the previous state). Recurrent neural network (RNN) is a type of neural network that resolves the problem. In RNN the output from previous step is fed as input to the current step [21,22]. RNN solves this issue using a specially designed hidden layer which remembers some information about a sequence.

In RNN, the current state is calculated as the function of the previous state and the input:

$$h(t) = f(h(t-1), x(t))$$

where $h(t)$ = current state

$h(t-1)$ = previous state

$x(t)$ = input state

The function 'f' can be a nonlinear activation function such as tanh. The output $y(t)$ is calculated as:

$$y(t) = W \cdot h(t)$$

where W is the weight at the output layer.

In order to train RNN, a single time step of the input is first provided to the network. Next, the current state is calculated using the current input and the previous state. Then the current state for the next time step is calculated. This can go as many time steps as possible according to the problem and join the information from all the previous states. Once all the time steps are completed, the final current state is used to calculate the output. The output is then compared to the actual target output to generate the loss function. The error is then backpropagated to the network to update the weights of the network.

RNNs are used in many applications such as predicting the next word from a text sequence, natural language processing (NLP), and also for classification purposes. It can also be used in time-series analysis in biomedical signals including ECG. A typical example is the prediction of the next AF beat from an ECG.

RNNs have some limitations: It cannot deal with very long sequences. There are advanced deep learning algorithms such as gated recurrent unit

(GRU) [23], and long short-term memory network (LSTM) [24] that resolves the problems of RNN.

3.7 Semisupervised learning in biomedical signal processing

Up until this point, we have used ML as a tool to classify two or multiple different classes. In all these cases we use the supervised learning approach. We need a training phase in order to teach the classifier about the boundary of the two classes. The training instances need to be labeled in the training process. Now, consider a scenario where we want to classify normal and abnormal heart sounds from digitally-recorded PCG signals. Abnormal heart sounds can have diverse morphological and spectral characteristics due to the underlying pathological conditions. For example, a person with a cardiac murmur may often have extra heart sounds in between the fundamental heart sounds S1 and S2. Patients who have coronary artery disease and valvular disease often show other types of signatures in their heart sound. In order to train an optimum binary classifier, we need to incorporate all types of abnormal heart sounds in the in-training set. This is quite a difficult task in a practical scenario as there are thousands of pathological conditions that may end up causing an abnormal heart sound. On the contrary, there are some well-known properties that should be present in a normal heart sound. For example, a PCG signal recorded from a healthy normal person should have a dominant S1 and S2 heart sounds in every cardiac cycle. The spectrum of the fundamental heart sounds should also contain its energy within a given frequency region. Normalcy can also be determined from the mean heart rate, which should be within the range of 60–90 bpm. Failing to meet this condition can be considered in a PCG signal as an abnormal recording. Suppose we train a classifier to learn the properties of the normal heart sound. It sounds like a feasible approach as normal PCG signals are easily available. During testing, if the classifier finds that the properties of the test instance closely match with the properties of the training data, the test instance is marked as normal, otherwise it is predicted as an abnormal recording [25].

An autoencoder is a popular semisupervised approach that is used in the field of biomedical signal processing for determining abnormal segments, which are known as an anomaly [26].

A simplified block diagram of an autoencoder is shown in Fig. 13. It has two components: An encoder and a decoder. The encoder takes the input and converts it to an intermediate encoded vector which is of much smaller

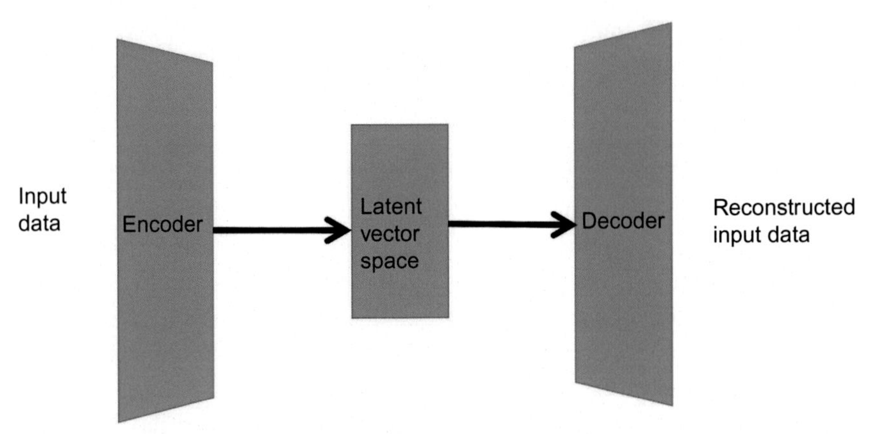

Fig. 13 Simplified block diagram of autoencoder.

dimension than the original training data. This representation is called the latent vector. The decoder performs the inverse task, i.e., it tries to reconstruct the original training data from the latent space. Both the encoder and decoder are designed using nonlinear functions to undertake the corresponding mapping. The encoder and decoder can employ complex deep learning structures such as CNN as well. The objective of an autoencoder is to effectively reconstruct back the original data from the encoded feature space, i.e., the mean squared error between the original and the reconstructed data should be minimized. During training, an autoencoder tries to minimize the difference between the original and the reconstructed training data by modifying the weights in the corresponding encoder and the decoder blocks via backpropagation. The mathematical difference between the two is termed the reconstruction loss. When the training is finished by minimizing the reconstruction loss, the latent vector can be considered an efficient encoded representation of the input training data in a much reduced dimension. Now, the obvious question arises regarding the utility of a network that generates its own input. The utility lies in the encoded space that contains the important information of the original data in a much reduced space. This feature of an autoencoder enables it to be used in a number of practical applications including data compression and anomaly detection.

Now we will discuss how autoencoders can be used for anomaly detection. Let's go back to our original problem statement of detecting normal and abnormal heart sounds. We can assume that normal and abnormal heart sounds are mutually exclusive and collectively exhaustive. That means features of normal and abnormal heart sounds are disjointed and they together complete the entire set of heart sounds. However, normal heart sounds have a known spectral property with spectral energies concentrated at certain known regions typically below 100 Hz. Heart sounds that do not meet that property can be marked as annotated. We design a classifier based on semisupervised learning where we only need annotated normal heart sound data for training. The autoencoder may have the spectrogram of the heart sounds as the input and its encoder and decoder may contain CNN as the building block. Once the network is trained, it is expected that it can learn the pattern of normal heart sounds very well and can encode them. During testing, if an unknown heart sound comes in as input, the reconstruction loss of the autoencoder can be considered as a parameter to determine whether it is normal or abnormal. We term this parameter the anomaly score. If the autoencoder is trained well with substantial normal data, the encoder should be successfully able to encode any normal data to the latent space and the decoder should be able to reconstruct the original data from the encoded space with a minimum reconstruction loss. If an unknown PCG recording after undergoing the pretrained autoencoder reports a very small value of reconstruction loss to regenerate the original data, it is predicted as normal, otherwise it is predicted as an anomaly. The threshold value of the anomaly score for the decision making needs to be effectively chosen for optimum performance.

4. Further challenges

So far, we have covered the basic steps needed to analyze sensor signals. In the following subsections, we will cover certain advanced processing steps that are gradually becoming part of clinical data analysis.

4.1 Handling unbalanced data

The quality of a supervised ML technique depends on multiple parameters such as the amount of data, generally the more the merrier. However, with an increasing number of classes, it is generally more difficult to classify them. In multiclass classification, it is important to learn enough discriminating features for each of those classes against the rest of the classes—and this requires enough representation of each class. Hence the question becomes, how much is enough? Is there any rule of thumb?

In this subsection we will cover the unbalanced data problem and demonstrate why it is more important in multiclass multilabel scenarios.

Generally, an unbalanced dataset problem is handled in one of the following manners:

1. Undersampling of the majority class: If the dataset has enough representation of the minority class, then undersampling the majority class in order to bring a parity amongst minority and majority representation is the way to go. This is easy to implement, however the major drawback is that a part of the data representing the majority class is dropped and the learner fails to learn enough features.
2. Oversampling minority class: If the minority class does not have enough data, then one possibility is to oversample the minority class by randomly picking up minority instances multiple times. Although this is easy to do, and ensures that no data is left behind, the disadvantage is overfitting due to duplicated data. Remember that the duplication can be manifold as it depends on the scarcity of the minority data.
3. Synthetic data generation: In 2002, Chawla et al. introduced the very popular synthetic minority oversampling technique (SMOTE). Since then, multiple extended flavors of SMOTE have been presented. The adaptive synthetic sampling approach (ADASYN), proposed by He et al. in 2008, is also a very popular technique to generate synthetic minority data. However, it must be noted that although SMOTE and ADASYN have advantages in various classification problems, they are yet to be found very effective on clinical data.

Classification problems can be broadly categorized based on the count of classes and labels as follows:

- Binary classification: This is the simplest form of classification problem where the dataset contains data from two classes.
- Multiclass: When the classification problem involves more than two classes, it is termed a multiclass classification problem.
- Multilabel: This means class memberships are not mutually exclusive. One instance can be annotated as more than one label—hence the phrase multiple label classification.
- Multilabel multiclass (MLMC): This is the toughest nature of a classification dataset. This means that the problem has multiple classes and each of the instances can have multiple labels. In mathematical terms, this has huge implications. For example, we may be dealing with a dataset of 100 classes, which is a realistic number in cardiac dataset (see the discussion of 12-lead ECGs in the 2020 PhysioNet Dataset referred to in Chapter 1). Now let us calculate the possible number of unique set of labels. How many unique possibilities of 1 label are possible? The answer is ${}^{100}C_1$.

How many unique possibilities for 2 labels? That is $^{100}C_2$, and so on. Hence, the total number of possible unique labels end up being $\sum_{k=1}^{100}\binom{100}{k}$

No amount of data in the universe will be enough to provide sizable representation of each of those unique possibilities. As a matter of fact, there will be patients having a rare combination of cardiac diseases (i.e., labels) and no amount of data collection will be nearly enough. As this type of detailed multilabel annotation in a cardiac dataset is very new in nature, the preparation of data for a MLMC classification is still a research problem and we will show how to deal with that in the next section (see also Chapter 4).

4.2 Clinical knowledge segmentation

The annotation of classification problem is undertaken by the automated observations (such as machine failures) or by domain experts. In any clinical classification dataset, the annotation is always performed by medical experts. This is a tricky situation as not only do the experts bring in their years of continuing experience during annotation, but they also greatly differ in opinion. In technical terms, this is expressed as interrater reliability. More often than not, experts differ in opinion. The degree of difference, i.e., interrater reliability, varies from one use case to another. For example, marking atrial fibrillation has a higher interrater reliability when compared to other lesser-known arrhythmias. In sleep analysis, the annotator needs to annotate an entire sleep session of the user (typically ranging from 7 to 10 h). Each user is attached to 12–24 different sensors. As one can presume, this is a very demanding task of rating each minute of approximately 8 h of data where the annotator needs to catch certain interrelation of those 12–24 signals. As a result, the interrater agreement amongst sleep annotators is as low as 70%. Hannun et al. [27] went one step further and claimed that the proposed deep network is able to catch different types of arrhythmias better than an average cardiologist. They substantiated their claim by having multiple annotations for each instance and then creating a more robust ground truth from this team of doctors. They trained with that robust annotation and successfully beat individual cardiologists' performances when measured against the robust ground truth.

However, whether ML will replace a doctor or not is a long debate. For now, we subscribe to the notion that a doctor who uses AI is better equipped than a doctor who does not. Similarly, a ML approach which uses the huge domain knowledge of the medical fraternity is also better than an approach which does not. Hence, we foresee that clinical classification problems, especially cardiac ones, will be better managed by a hybrid approach. A good starting point can be the marriage of traditional features (extracted as per the inputs of medical knowledge) and a deep architecture (which utilizes the enormous calculation capability of deep learning architecture which is beyond human realms).

5. Conclusion

Biomedical signal processing is the core element of digital healthcare. Biomedical signals can often be extracted in terms of digital time series. The time series contains important information that can be used for identification of cardiac diseases using digital techniques including AI and ML approaches. In this chapter, we have discussed the process flow for analyzing biomedical data to process meaningful information. Digitally recorded biomedical data is vulnerable to background noise. Hence, special care needs to be taken before processing. We have described a set of operations that can be used for noise cleaning of biomedical signals. Then we discussed how ML and deep learning-based techniques are designed on biomedical signals. We have listed a few popular features extracted from PPG, PCG, and ECG that are used in various practical operations. We have also briefly covered how ML and deep learning algorithms can be used for disease classification. One must remember that feature extraction is the most important part of a classification task. The desired classification may not be achieved if the features are not calculated correctly. In this chapter, we have given a brief idea regarding the generic process flow of biomedical signal analysis. In subsequent chapters, we will discuss how the process can be modified for use in more practical biomedical applications.

References

[1] T. Bhattacharjee, A.D. Choudhury, A. Pal, Robust beat-to-beat interval from wearable PPG using RLS and SSA, in: 2019 41st Annual International Conference of the IEEE Engineering in Medicine and Biology Society (EMBC), IEEE, 2019, pp. 4946–4952.
[2] Z. Reitermanova, Data splitting, in: WDS, vol. 10, 2010.
[3] J. Allen, Photoplethysmography and its application in clinical physiological measurement, Physiol. Meas. 28 (3) (2007) R1.
[4] S. Karpagachelvi, M. Arthanari, M. Sivakumar, ECG Feature Extraction Techniques—A Survey Approach, arXiv preprint arXiv:1005.0957, 2010.
[5] M. Kyoso, A. Uchiyama, Development of an ECG identification system, in: 2001 Conference Proceedings of the 23rd Annual International Conference of the IEEE Engineering in Medicine and Biology Society, vol. 4, IEEE, 2001.

[6] C. Liu, et al., An open access database for the evaluation of heart sound algorithms, Physiol. Meas. 37 (12) (2016) 2181.
[7] S. Wold, K. Esbensen, P. Geladi, Principal component analysis, Chemom. Intell. Lab. Syst. 2 (1–3) (1987) 37–52.
[8] H. Abdi, L.J. Williams, Principal component analysis, Wiley Interdiscip. Rev. Comput. Stat. 2 (4) (2010) 433–459.
[9] P. Comon, Independent component analysis, a new concept? Signal Process. 36 (3) (1994) 287–314.
[10] J.V. Stone, Independent Component Analysis: A Tutorial Introduction, The MIT Press, 2004.
[11] R.E. Wright, Logistic Regression, Open Journal of Leadership, 1995.
[12] D.G. Kleinbaum, K. Dietz, M. Gail, M. Klein, M. Klein, Logistic Regression, Springer-Verlag, New York, 2002.
[13] W.S. Noble, What is a support vector machine? Nat. Biotechnol. 24 (12) (2006) 1565–1567.
[14] J. Platt, Probabilistic outputs for SVMs and comparisons to regularized likehood methods, in: Advances in Large Margin Classifiers, MIT Press, 1999.
[15] A.J. Myles, R.N. Feudale, Y. Liu, N.A. Woody, S.D. Brown, An introduction to decision tree modeling, J. Chemometr. 18 (6) (2004) 275–285.
[16] L. Breiman, Random forests, Mach. Learn. 45 (1) (2001) 5–32.
[17] Y. LeCun, Y. Bengio, G. Hinton, Deep learning, Nature 521 (7553) (2015) 436–444.
[18] I. Goodfellow, et al., Deep Learning, vol. 1. no. 2, MIT Press, Cambridge, 2016.
[19] R. Rojas, The Backpropagation Algorithm. In Neural Networks, Springer, Berlin, Heidelberg, 1996, pp. 149–182.
[20] P. Baldi, P.J. Sadowski, Understanding dropout, Adv. Neural Inf. Proces. Syst. 26 (2013).
[21] L.R. Medsker, L.C. Jain, Recurrent neural networks, Des. Appl. 5 (2001) 64–67.
[22] T. Mikolov, M. Karafiát, L. Burget, J. Cernocký, S. Khudanpur, Recurrent neural network based language model, in: Interspeech, vol. 2, 2010, pp. 1045–1048. no. 3.
[23] J. Chung, C. Gulcehre, K. Cho, Y. Bengio, Empirical Evaluation of Gated Recurrent Neural Networks on Sequence Modeling, 2014. arXiv preprint arXiv:1412.3555.
[24] S. Hochreiter, J. Schmidhuber, Long short-term memory, Neural Comput. 9 (8) (1997) 1735–1780.
[25] X. Zhu, A.B. Goldberg, Introduction to semi-supervised learning, Synth. Lect. Artif. Intell. Mach. Learn. 3 (1) (2009) 1–130.
[26] D. Bank, N. Koenigstein, R. Giryes, Autoencoders, arXiv preprint arXiv:2003.05991, 2020.
[27] A.Y. Hannun, P. Rajpurkar, M. Haghpanahi, G.H. Tison, C. Bourn, M.P. Turakhia, A.Y. Ng, Cardiologist-level arrhythmia detection and classification in ambulatory electrocardiograms using a deep neural network, Nat. Med. 25 (1) (2019) 65–69.

SECTION 2

Disease screening

CHAPTER 4

Abnormal heart rhythms

1. Introduction

From an engineer's point of view, the human heart is a mechanical pump with four chambers: Left and right atria and left and right ventricles. This pump circulates blood with a rhythm determined by sinus nodes (also known as sinoatrial nodes or SA nodes). It receives deoxygenated blood which flows into the right atrium through the superior and inferior vena cava. It is then pumped into the right ventricle, followed by the lungs, where the exchange of oxygen and carbon dioxide takes place. This process is commonly known as pulmonary circulation. The oxygenated blood comes back to the left atrium, followed by the left ventricle, and is finally ejected into the entire body; this is popularly known as systemic circulation. During systemic circulation, again, the exchange of oxygen and carbon dioxide takes place in the blood capillaries throughout the body. After this exchange, the deoxygenated blood comes back to the right atrium. This entire series of events, blood leaving left atrium to blood flowing back to right atrium, as shown in Fig. 1, is known as one cardiac cycle.

So what causes the mechanical pumping actions of the human heart? There is an underlying electrical conduction system which controls and coordinates these four chambers. At the top of the heart, there exists a collection of tissue commonly known as the SA node. This is the pacemaker of the human heart. In a normal human heart, the SA node will generate regular electrical stimulus, roughly 60–100 times per minute. Once the SA node is fired, the upper heart chambers, i.e., left and right, are activated. In the process, both the atria contract and push the blood to the left and right ventricles. The electric stimulus approaches downwards though the conduction pathways and reaches the atrioventricular node (AV node) from SA node. The AV node is situated between the atria and ventricles. It slows the electrical signal, thereby allowing sufficient time for the ventricles to receive blood from the atria. Subsequently, the AV nodes send the electrical impulse to a bundle of conduction cells known as the bundle of His, which divides into two conduction pathways called the left and right bundle

New Frontiers of Cardiovascular Screening using Unobtrusive Sensors, AI, and IoT
https://doi.org/10.1016/B978-0-12-824499-9.00004-0

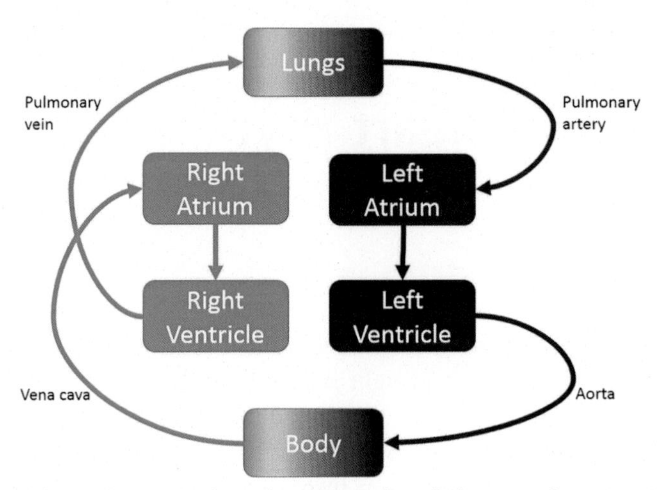

Fig. 1 Blood flow in human body (*red and blue colors, dark gray and gray in print version,* depict oxygenated and deoxygenated blood respectively).

branches. These two branches divide into conducting fibers which propagate the signal into the ventricles, and the ventricles start contracting. The right ventricle pumps blood to the lungs and the left ventricle pumps it to the rest of the body. When the SA node sends another signal to the atria, the next cardiac cycle starts.

Any deviation from this electrical flow may result in abnormal rhythms. Before delving into the topic of abnormality, let us first look at normal rhythm, and how it may vary. The heart rate is the basic parameter to measure cardiac rhythm. Heart rate is commonly measured as the number of heartbeats taking place in 1 min. The global standard resting heart rate for adult humans is 72 beats per minute (bpm). A heart rate can vary depending on a lot of factors ranging from age, physical fitness, genetics, environment, diet, or medication. When a person starts some form of exercise, for example running, walking, or even going from a sleeping position to a standing position, the heart rate increases to provide more oxygenated blood. A higher physical fitness level tends to correlate with lower resting heart rate, e.g., athletes tend to have lower heart rates (can be as low as ~50 bpm). This can be explained by the inherent high cardiac output (CO) they develop during the process of years of rigorous cardiac training. CO is defined as the liters of blood pumped by a human heart rate per minute. For a normal sized adult human, the typical CO is 5 lpm, and this is generally higher for an athlete [1]. As a result, an athlete's heart needs fewer beats per minute compared to a nonathlete.

If the normal sequence of electrical impulses change, it results in the deviation of the speed and rhythm of the heart beats, medically termed arrhythmia. The mechanisms leading to cardiac arrhythmias can be broadly attributed to two categories: Abnormal impulse and conduction disturbances [2]. The term arrhythmia covers many different conditions, all of which are different forms of irregularities in heart beats. A normal adult heart rate can vary between 60 and 100 bpm. If the heart is beating too slow, i.e., below 60 bpm, it is called bradycardia. When symptomatic, it may cause dizziness, fatigue, and even fainting. On the other hand, if the resting heart rate goes past 100 bpm, it is known as tachycardia. With an abnormally high heart rate for a sustained period of time, a person with tachycardia generates more internal friction, leading to unwanted turbulence and swirling in the blood while flowing through the blood vessels. Arrhythmias are a set of cardiac conditions, each of them representing a form of rhythmic irregularities of the human heart. Some of the other commonly known arrhythmias are:

1. Supraventricular arrhythmias:
   a. Atrial fibrillation: More commonly known as AFib or AF, this is one of the most common arrhythmia. In AF, the upper chambers of the heart start a rapid and irregular rhythm. Heart beats become irregular. AF may lead to fatigue, shortness of breath, or palpitation. It also increases the chances of a heart stroke. Studies have found that 1 out of 7 strokes are attributed to AF [3].
   b. Atrial flutter: This condition occurs when a short circuit in atria leads to very frequent pumping (flutter) of the upper chambers. Typically, 240–340 contractions take place in a minute. As a result of these rapid contractions, the atria do not have enough time to be filled with blood. Atrial flutter often causes blood clots, leading to the obstruction of blood vessels, a fatal condition known as thromboembolism [4].
2. Ventricular arrhythmias:
   a. Premature ventricular contraction (PVC) and ventricular bigeminy: In this type of arrhythmia, there are extra heartbeats that originate in the ventricles [5]. PVCs are common and singular PVC events generally do not pose a serious threat [6]. However, frequent or regular PVCs can lead to serious conditions. If there is a single irregular heartbeat occurring after every normal heartbeat, then that PVC is called ventricular bigeminy.
   b. Ventricular tachycardia (VT) and ventricular fibrillation (VFib): During VT, the lower chambers of the heart go into fast regular beats. If

VT persists for a long time, it may lead to other more severe arrhythmia such as VFib [7]. During VFib, the lower chambers go into fast irregular beating.

3. Heart blocks [8]:
   a. First-degree: Electrical impulses reach from atria to the ventricles through the AV node in a slower fashion. This is the mildest form of heart block. If not treated, this may progress into more serious forms such as second-degree and third-degree heart blocks.
   b. Second-degree: During this arrhythmia, one or more impulses fail to reach the ventricles from the atria. Type 1 second-degree blocks, also known as Wenckebach, is the less serious form where heart beats get slower until it skips a beat. In type 2 second-degree block, more heart beats tend to be skipped.
   c. Third-degree: This is the most severe form of heart blocks where none of the impulses succeed in reaching the ventricles from the atria. This condition is also commonly referred to as a complete heart block.

Each of the arrhythmias mentioned above show up as certain patterns in the corresponding ECG. On the other hand, for simpler medical parameters such as heart rate or heart rate variability, we can work with either photoplethysmogram (PPG) or ECG. In this chapter, we will review how to prepare common time-series sensor signals such as PPG and ECG; and detect heart rhythm abnormalities using machine learning (ML) and other signal processing techniques.

2. Heart rate measurement using PPG and ECG

A heart rate can be measured in a variety of ways. The simplest method is to place one's index and middle fingers on the back side of the wrist, 1–2 in. away from the palm. Once placed, exert slight pressure until you feel pulsing. One may have to move the fingers a little bit up and down to find the pulse. Another common way to check a pulse is to place one's fingers on the carotid artery (on the neck by the side of the windpipe). However, all these methods have a few issues: (1) Certain expertise is required to find the pulse, even more if one wants to find out abnormality in pulses, (2) Cannot be used as a 24×7 method to monitor and look for abnormal episodes, and most importantly, (3) Complex rhythm abnormalities, which depend on the morphology of the physiological signals, cannot be determined from heart values alone.

Heart rate abnormality analysis using PPG or ECG removes these issues. At the time of writing, the consumer market is flooded with lifestyle low-cost devices with PPG and ECG sensors. The trend started with smartphone accessories in between 2010 and 2015, and between 2015 and 2020, these two sensors have become a must-have for smartwatches [9,10]. Given the sharp rise in the volume of these devices, and ever-lowering price, continuous heart rate measurement is all set to become a de facto standard. Every other week or month, a company is receiving US Food and Drug Administration (FDA) approval of their medical-grade smartwatch, which further strengthens the fact that the accuracy and precision of these devices have been elevated from toy devices to serious medical-grade devices. Last but not least, all these devices are connected to Internet, providing real-time alerts not only to the subjects but to the first responders, physicians, and other care givers. Given this huge potential, let us take a step back and look at the basics of ECG and PPG signals and check how to process them.

2.1 Preprocessing: Smartphone, wearable, and nearable

It can be noted that the position of PPG and ECG sensors used in smartwatches and smartphone accessories are bit unorthodox. In a smartwatch, both the sensors are typically attached to the wrist band. In a smartphone setup, the PPG is collected from the video of the fingertip with the flashlight on. In order to collect the ECG, generally an accessory in the form factor of a phone cover is used, where the user needs to place the fingers on the metal leads of the nominated accessory. In the past decade, with the ubiquitous presence of cameras in laptops, desktops, phones, and work environment, face PPG has emerged as a new source of PPG [11]. PPGs and ECGs collected in this way contain much more noise as opposed to the data collected by a medical grade PPG or ECG device under a stricter protocol such as rest position. The source of the noise is mainly artifacts introduced through physical movements. For example, when someone wearing a smartwatch is moving, especially in a chaotic, nonrepetitive manner, say typing on a keyboard, or casual walking or driving. All these activities insert unwanted disruption in the morphology of the signal. Depending on the level of noise, the sensor signal may be incorrigible or otherwise. It is important to throw away the incorrigible signal as processing would lead to unreliable conclusions. The following are the most common processing steps for processing raw signals [12]:

1. Given that the human heart rate ranges between 50 bpm and 200 bpm, one can create a bandpass filter with a pass band of 0.5–5 Hz (30–300 bpm). Any low frequency (<0.5 Hz) noise such as slow-motion artifacts or high frequency noise such as jittery hand movement will be removed from the signal. However, any noise within the pass band range would still very much reside in the signal and may lead to erroneous calculation of heart rate. As discussed in Chapter 3, there are a handful of digital filters available. However, one of the most popular choices is a 2nd order or 4th order Butterworth Bandpass filter.
2. The next step is to normalize the PPG signal. There are numerous options for normalization such as linear scaling, clipping, or log-scaling. However, for the problem at hand, one of the most efficient and popular techniques is normalization to zero mean and unit variance, also known as z-score normalization. Normalized value $z=(x-\mu)/\sigma$, where x is the raw value, μ is the mean, and σ is the standard deviation.
3. The human heart rate, even at the abnormal end, is not very high frequency. We have assumed the highest possible heart rate as 300 bpm (5 Hz). As per the Nyquist Theorem, the sampling rate of the PPG or ECG should be at least twice of that, i.e., 2×5 Hz $=$ 10 Hz. Hence, if the data at hand has a higher sampling rate, we can downsample it to a lower sampling rate, such as 20—-25 Hz.
4. The next step in the processing pipeline is the denoising. Up to now we have removed the artifacts which are outside our interest of frequency (0.5–5 Hz). However, there are motion artifacts (MA) present in this frequency range too and several signal processing tools can be deployed to help. We will cover these methods in the next section.

2.2 Frequency and time domain analysis

PPG and ECG signals without any MAs can be analyzed in both a time and frequency domain. In the time domain, the period of a cycle of these signals represents the cardiac cycle. In the frequency domain, the most prominent peak in the fast Fourier transform (FFT) should indicate the average heart rate of the person [13]. However, this process has a few major disadvantages:

1. Oftentimes there are spurious peaks, especially if it is a diseased PPG or ECG. For example, a person with AF will have an irregular heart rate. The cardiac intervals will be completely chaotic. In the case of other

arrhythmias, where a subject may have multiple dominant rhythms, multiple such peaks will appear in the frequency domain and none of them will singularly present the subject's condition.

2. Frequency domain analysis will not give us beat-to-beat cardiac interval (the inverse of instantaneous heart rate), which is the basic requirement to detect any cardiac rhythm anomaly.
3. Depending on the sampling rate and data length of the signal in question, the error of heart rate calculation would significantly vary. For example, if one algorithm processes 10 s of data in the frequency domain to find the average heart rate, versus the same technique applied on 20 s of data, the second approach would tend to provide overall fewer errors in multiple tries as the multiplication factor would be decreased. However, this will come at a cost of waiting for double data collection. Also the chances of recording clean signal for 20 s is less than that of 10 s.

That leads us to the world of wearables, where the major challenge is to remove MAs from ECG and PPG signals. The IEEE Signal Processing Cup (SPC) 2015 provided such a challenge. This wrist PPG SPC dataset contained a two-channel wrist PPG along with three-axis accelerometer and single channel chest ECG data [14]. The subjects ran on treadmill with a predefined protocol involving rest, brief walking (6–8 km/h), and running (12–15 km/h). As the ECG signal was collected using chest-strapped sensors, it was almost noise free and provided the true heart rate. In addition to the free swing of hands during running, the subjects were encouraged to purposefully use their hand with the wristband for other activities such as pull clothes, wipe sweat, push buttons on the treadmill, etc. This made the wrist PPG signal contaminated with substantial MA. The challenge was to use PPG and accelerometer signals to estimate the heart rate. The errors were calculated against the heart rate estimates calculated from the ECG. Researchers used different signal processing techniques to remove the MAs in the PPG signal of wrist PPG SPC dataset. Table 1 summarizes some of the successful techniques. Most of the approaches were able to reach answers within 1–2 bpm, which is accurate enough to track the real-time heart rate. Also, one can note that the approaches that used an acceleration signal to detect the MAs were more successful than the rest. Very few approaches employed supervised deep learning network, probably owing to the smaller size of the wrist PPG SPC dataset.

Table 1 State-of-the-art methods to remove motion artifact on wrist PPG SPC dataset.

Input signal	Brief methodology	Result
PPG; acceleration	Signal decomposition is performed followed by sparse signal reconstruction using high-resolution spectrum estimation, and spectral peak tracking is used to estimate heart rate [14].	2.42 bpm error
PPG; acceleration	Multiple measurement vector (MMV) is applied in order to recover a joint sparse signal, common spectral features of PPG and acceleration. MA frequency bins are removed using spectral subtraction on the reconstructed spectrum [15].	1.28 bpm error
PPG	Window-based adaptive noise cancellation in order to remove in-band noise. Finally, heart rate is estimated using multiinitialization spectral peak tracking [16].	1.28 bpm error
PPG; acceleration	Enhanced Wiener filtering is used to reduce MAs using noise patterns from acceleration. A phase vocoder is used to further refine heart rate estimates [17].	1.02 bpm error
PPG; acceleration	A spectral filter is designed using power spectral density of both the signals [18].	0.99 bpm error
PPG; acceleration	A Wiener filter is used to denoise PPG. A finite state machine (FSM) is proposed as postprocessing step [19].	1.25 bpm error
PPG; acceleration	Singular value decomposition (SVD) of acceleration along with adaptive filtering is used. Then an iterative adaptive thresholding is applied to generate the higher resolution spectrum of the processed signals [20].	1.25 bpm error
PPG; acceleration	Ensemble empirical mode decomposition (EEMD) is used, followed by recursive least square (RLS) filters to refine the final heart rate estimates [21].	1.02 bpm error
PPG	Short term Fourier transform (STFT) and spectral analysis were employed to estimate heart rate [22].	1.06 bpm error
PPG	A supervised deep learning framework using two-layer CNN and two-layer LSTM were used to estimate heart rate [23].	1.47 bpm error
PPG; acceleration	Singular spectrum analysis (SSA), followed by adaptive RLS is used to isolate the oscillatory components and noise. Advanced weighted local interpolation followed by polynomial fitting and an outlier removal step were used for peak correction [24].	1.68% error

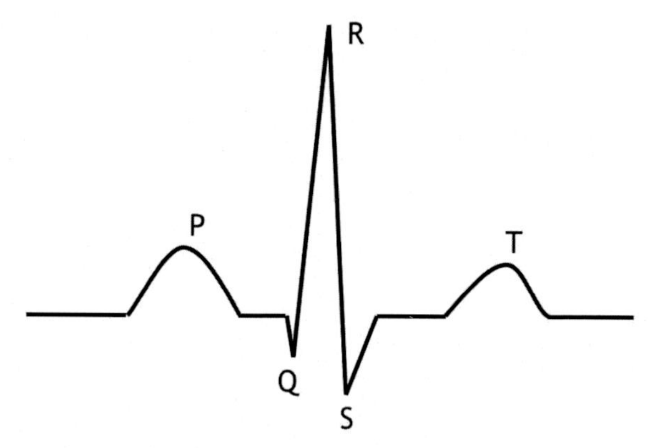

Fig. 2 ECG features.

3. Arrhythmia detection using PPG and ECG

Although the 12 leads of an ECG differ in morphology, they have some common features as shown in Fig. 2. The P wave, caused by depolarization of the atria, precedes the QRS complex. The QRS complex is a high amplitude spike with narrow base of 80–100 ms, caused by rapid depolarization of the ventricles. The PR interval, the distance from the beginning of the P wave to the beginning of the QRS complex, is an important marker. A longer PR interval has a high correlation with cardiac diseases such as first-degree blocks. Similarly, if the width of QRS complex is wider than 120 ms, it may signify left and right bundle branch blocks (LBBB and RBBB). After the QRS complex, there is another wave called the T wave. A T wave along with a QRS complex gives rise to two important markers: The ST segment and the QT interval. An elevated or depressed ST has high correlation with ischemia. An elongated QT interval, also known as Long QT syndrome (LQTS), indicates the possibility of fainting, which may even lead to sudden death. It is to be noted that these are some of the examples of ECG anomalies, and most of them are often asymptomatic. Hence, early detection through automation is required.

3.1 Signal conditioning

Signal conditioning while detecting arrhythmia such as AF is tricky, as AF often looks like noise. Hence the aim here is not only to differentiate between corrupted signals and clean signals, but also detect the difference between abnormal rhythms and noisy signals. Given this problem definition, it is evident that a simple threshold on amplitude or signal energy will not be sufficient to differentiate noisy ECG and AF ECG. In order to solve this issue, researchers proposed an adaptive threshold which detects noise but lets the abnormal (yet clean) signal pass through [25]. To start with, short time Fourier transform (STFT) of raw ECG was performed. The normal sinus rhythm spectrogram has a regular comb-like morphology consisting of regularly spaced spikes. An ECG with AF will produce a comb structure with irregular spaces in between the spikes. Whereas in the noisy ECG segments, the teeth of the comb structure are blurred in the low frequency zone. Next, the sum of spectral power is computed for each STFT window which is then smoothened using the moving average. The smoothing window is intentionally kept larger than the STFT window size. This adaptive threshold will clear the clean segments (irrespective of them being normal sinus rhythm or AF) but it will raise an alarm for the noisy ECG segments. This approach provides an additional benefit as it points out the noisy segments instead of declaring the entire signal noisy. This provides us with the benefit of removing the noisy segments from an ECG strip and sending the rest of the signal for further processing. Please note that occurrence of noisy segments in handheld ECG devices is very common. In all probability, the subject just twitched a little bit for a second or two during those segments.

3.2 Feature engineering

ECG feature engineering can be broadly categorized into two methods. The first one detects QRS complexes in an ECG signal and the associated peaks (P, Q, R, S, and T), followed by extraction of features from individual cardiac cycles. This method creates a two-dimensional array of features where one dimension is the number of features and other dimension is the number of cardiac cycles of the corresponding specific ECG signal. Finally, some form of statistical aggregator (mean, median, kurtosis, or domain specific aggregators) are applied along the dimension of cardiac cycles in order to reduce it to a one-dimensional array of features whose length is equal to the number of the derived features (i.e., the number of statistical aggregators applied). The alternative approach is to consider the ECG signal as a whole and extract features directly from the entire signal without extracting the cardiac cycles.

The seminal Pan Tompkins algorithm and derivative of related research efforts produces extremely accurate (more than 95%) detection of R peaks in an ECG signal [26,27]. Also, the extraction of features from each cardiac cycle is closer to the conventional medical training of cardiologists. This is the reason the first method of feature extraction has been more popular. However, recent approaches have tended to take inspiration from both the approaches and make a hybrid solution.

Hundreds of features have been proposed in the past few decades. The following are the most important ones found discriminative in differentiating rhythm problems [27–43]:

- Skewness, kurtosis, peak count, maxima, minima of RR interval probability distribution
- Skewness, kurtosis, coefficient of variation, standard deviation of succeeding RR interval differences
- Mean of amplitude variation of RR peaks
- Count of succeeding RR interval difference being more than 20 and 50 milliseconds (commonly known as nn20 and nn50 in heart variability terminology)
- l_1 norm of RR and PP spectrum
- Shannon, Tsallis and Renyi entropies, 10th order linear predictor coefficients of ECG
- Spectral centroid, total harmonic distortion, zero crossing rate, kurtosis of ECG time series
- Ratio of R peak count and P peak count (P waves are generally not visible in ECG of subjects having AF)
- Normalized spectral area for frequency ranges such as 0–2 Hz, 2–4 Hz, 4–10 Hz, 10–150 Hz
- Coefficient of variation of QRS complex
- Median, range and coefficient of variation of S peak and R peak amplitude ratios, Q peak and R peak amplitude ratios
- Negative slope of ST segment normalized by QRS width
- Median of T peak and R peak amplitude ratios
- Median and coefficient of variations of QR width, QRS complex, RS slopes, ST slopes
- Median and Coefficient of variation of corrected QT segment
- Poincare plot geometrical features. Poincare plot is a two-dimensional plot, where the location of each point is RR_i, RR_{i+1}—for normal sinus rhythm all the points will be plotted in vicinity of each other, creating a very high density focus region at the center and a few outliers. Because AF demonstrates chaotic cardiac cycles, the respective Poincare plot would be completely random and haphazard in nature. For other rhythm problems, where there are multiple dominant rhythms, the Poincare plot may have multiple such focus regions.

- Bradycardia, tachycardia detection features (based on the average heart rate extracted from the RR intervals)
- Standard deviation and percentage of zero crossings of first order Intrinsic Mode Function (IMF)
- Lorenz plot features (AF evidence, origin count, irregularity evidence, pace evidence, density evidence, and anisotropy evidence) of RR intervals.

Detailing each feature mentioned above is beyond the scope of this book. However, one can use different strategies in ML while handling a multiclass problem. The PhysioNet 2017 Challenge presented such a competition where the challenge was to classify noisy handheld single-lead ECGs into one of the four classes: normal sinus rhythm, AF, other rhythm problems, and noisy signal [44]. The most naïve approach is to design a multiclass classifier with the features mentioned above. However, as one can imagine, not all the features will be discriminative for all classes. Researchers proposed a divide and conquer approach by employing a multilayer binary classification to mitigate this issue [25]. In this approach, in the first layer, a robust classifier was built with a subset of the features to segregate the ECGs into two buckets: {Normal and Other Rhythms} and {AF and Noisy signals}. This choice can be explained as AF signals often mimic the behavior of a noisy signal, both are visibly chaotic. On the other hand, for the normal rhythm and other rhythm classes, both had some intrinsic rhythmic patterns and hence sat in one bucket. It is to be noted that multiple feature selection techniques were employed to rank the features and only a few top-ranking features were selected in order to avoid the curse of dimensionality. The same method was repeated on the second layer, where two separate classifiers were built to distinguish {Normal and Other Rhythms} and {AF and Noisy signals} respectively. The proposed method generated 0.83 F1 score on the hidden test data.

4. Deep network for rhythm analysis

Let us look at the full spectrum of medical grade ECGs. An ECG is a chart of electric voltages against time. The electrodes are placed in different predefined contact points of the human body. The medical standard is a 12-lead ECG consisting of 3 limb leads (I, II, and III), 3 augmented limb leads (aVR, aVL, and aVF) and 6 precordial leads (V_1–V_6). In the past few years, a large

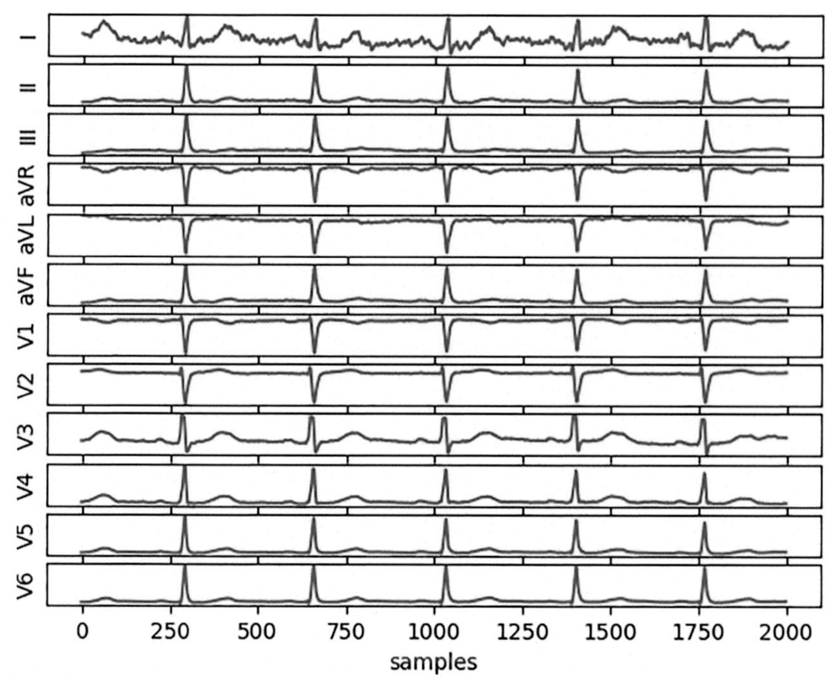

Fig. 3 12 lead ECG.

corpus of medical grade ECG equipment has become available in the private and public domains. Fig. 3 shows a q2 lead ECG record from PhysioNet 2020 dataset [45].

This type of huge dataset is not suitable for conventional path of signal cleaning or feature engineering followed by classifiers. First, the processing time to clean so many signals and extract features from each of them would be too time-consuming. Also, extracting domain-specific feature for each of those classes is a humongous task and impractical from a deployment perspective. Hence, deep neural network (DNN) is the most plausible way to tackle these types of huge datasets. Hannun et al. were the first to challenge the traditional ECG processing pipeline where they presented a DNN architecture tested on 91,000 single-lead ECGs with 12 different rhythm classes, collected from 53,000 subjects [46]. Their seminal work on automated arrhythmia detection produced performance measures comparable to trained cardiologists. Multiple cardiologists annotated the test data, and the mean interannotator agreement was rather low (72%). This highlights the scope of improvement of ground truth annotation in medical studies. In order to address this, they compared their performance with an average cardiologist—a cardiologist performance averaged across the six cardiologists who participated in this study.

The proposed architecture is a convolutional DNN which takes raw ECG data as input, and outputs a prediction for a predefined window of 1.28 s. The network has 34 layers and is motivated by the shortcut connection of residual network (ResNet) architecture. ResNET is a neural network which mimics the neurons of the cerebral cortex. They generally have multiple layers of rectified linear (ReLU) units and batch normalization. ResNet architecture is widely known for its efficacy toward mitigation of vanishing gradient problem. The proposed architecture also employs a dropout regularization technique to reduce overfitting. The obtained result revealed an excellent average area under the receiver operating characteristic curve (ROC) of 0.97. The average F1 score was also much better than that of the average cardiologist.

However, the proposed network does not take any patient information or traditional ECG features as input. This is certainly an area for improvement in future. One can introduce some of the traditional ECG features as well as patient information to create a hybrid approach, which may lead to a twofold improvement: Better performance and a more explainable pipeline.

The PhysioNet Challenge 2020 dataset is such an example, having more than 44,000 12-lead ECGs collected from the United States, EU, Russia, and China [45]. To the best of our knowledge, this is the largest public 12-lead ECG database available. Each data is labeled with one or more of the 27 classes, where each is a cardiac condition with a unique corresponding Systematized Nomenclature of Medicine Clinical Terms (SNOMED-CT). Figs. 4–20 demonstrate 17 such multilabel combinations. This makes the problem even harder as it becomes a multilabel multiclass (MLMC) classification problem. This means that the number of unique labels, both in theory as well as in practice, can easily reach thousands of unique labels. Additionally, a good percentage of those unique labels have a severely minor representation in the dataset, thereby making the whole classification problem extremely imbalanced. There are multiple ways to handle such balancing problems in ML. The most naive way is to go for oversampling and/or undersampling in order to reduce class imbalance as much as possible. However, there are disadvantages. Undersampling would leave a lot of samples out of training, whereas oversampling will bring in undesired overfitting to the training samples. Apart from 12-lead ECG time-series signals, there was some demographic information available such as age and gender. The labels were machine generated, reviewed by a single cardiologist, and adjudicated by a team of cardiologists. This was done to due to purify the

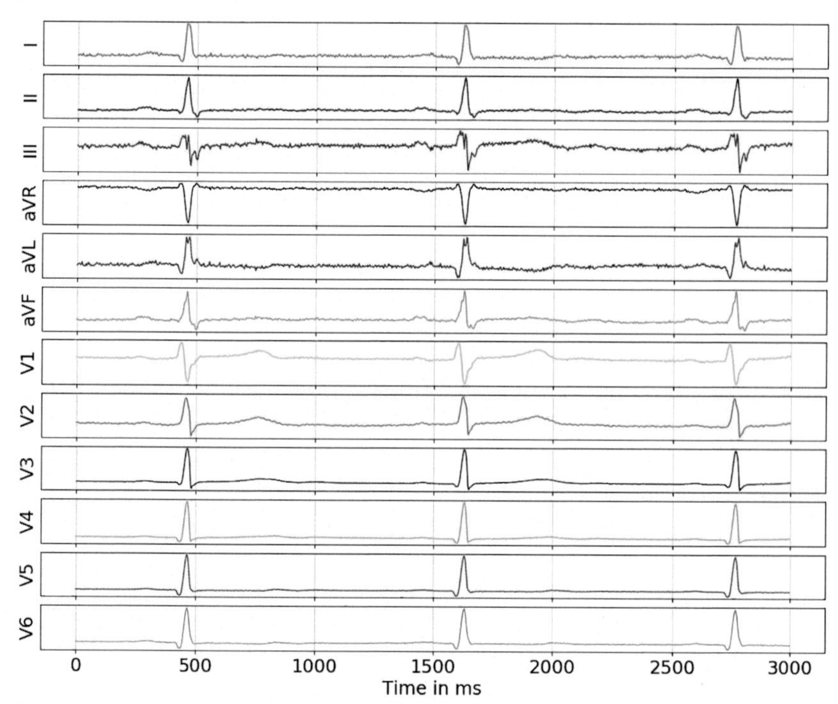

Fig. 4 12-lead ECG annotated as sinus bradycardia and abnormal T wave.

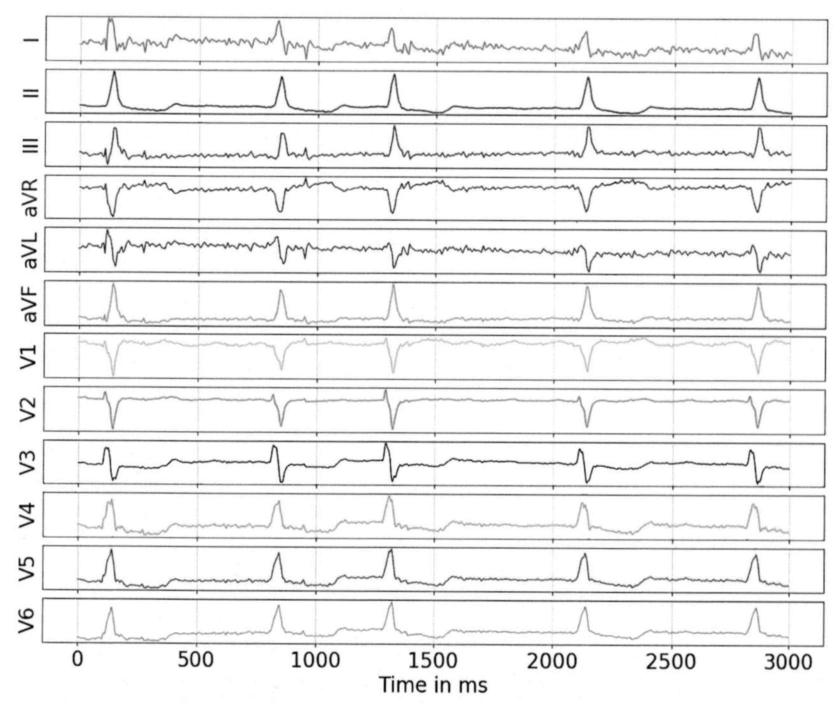

Fig. 5 12-lead ECG annotated as atrial fibrillation, abnormal T wave, and atrial flutter.

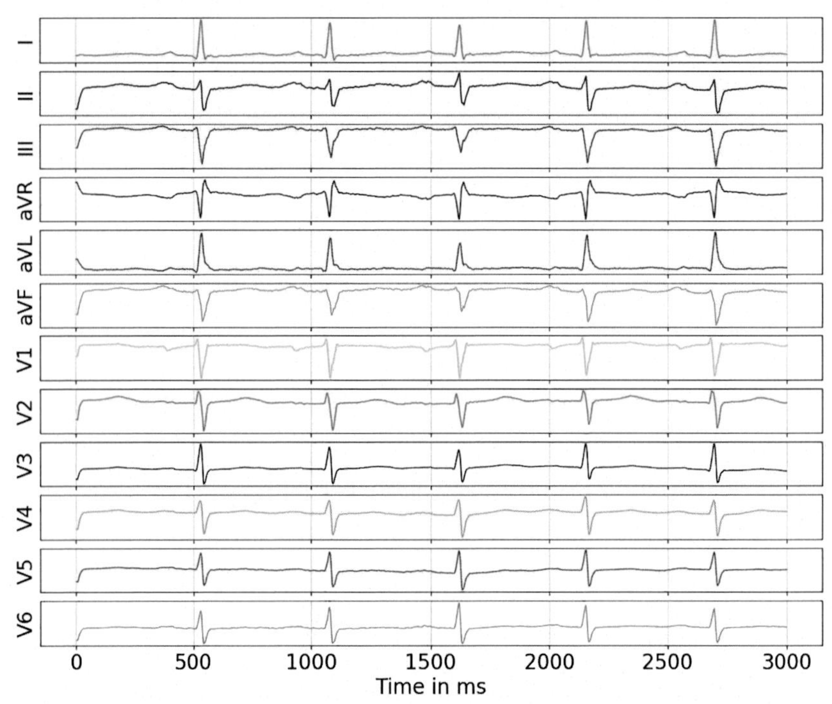

Fig. 6 12-lead ECG annotated as left axis deviation and sinus tachycardia.

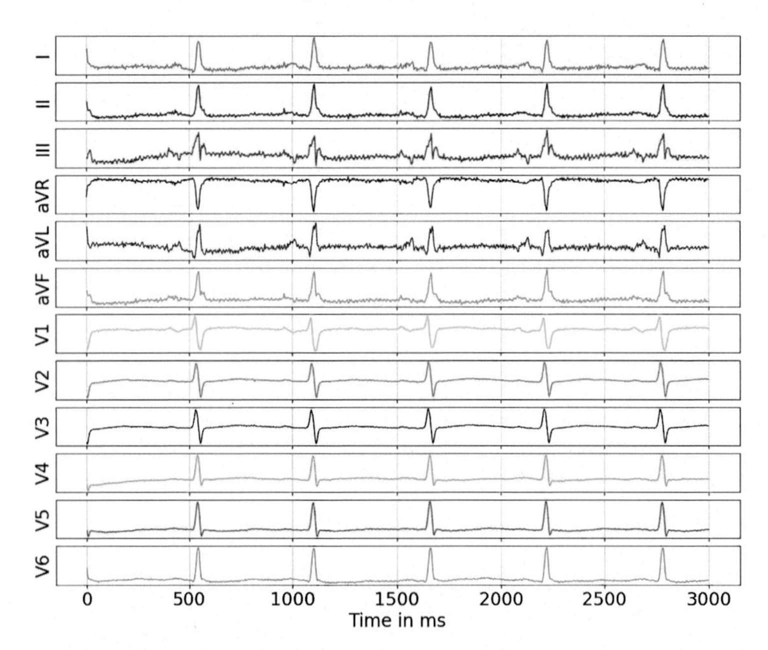

Fig. 7 12-lead ECG annotated as abnormal T wave, sinus tachycardia, and premature ventricular contractions.

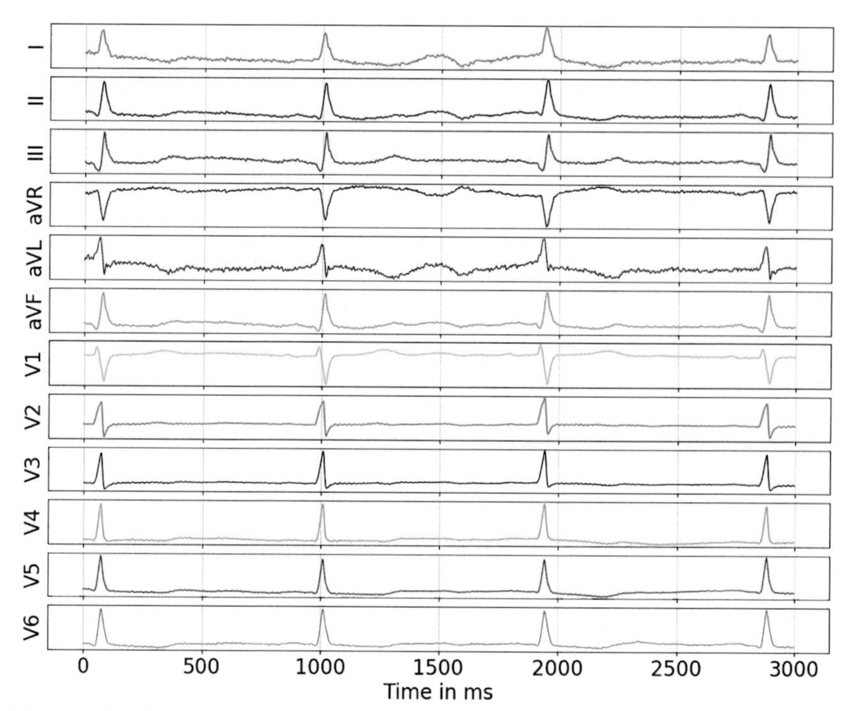

Fig. 8 12-lead ECG annotated as premature atrial contraction.

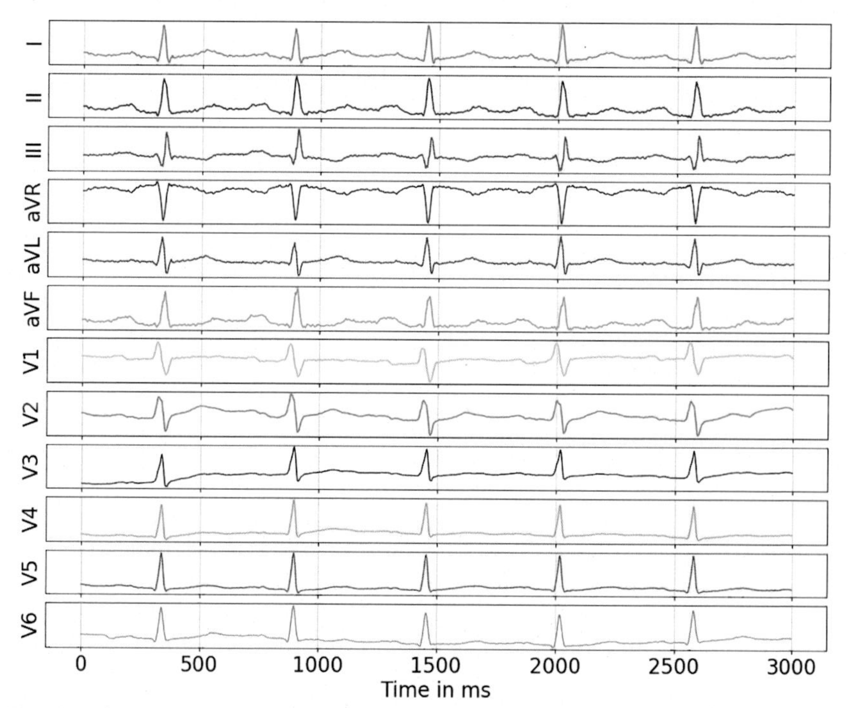

Fig. 9 12-lead ECG annotated as sinus tachycardia.

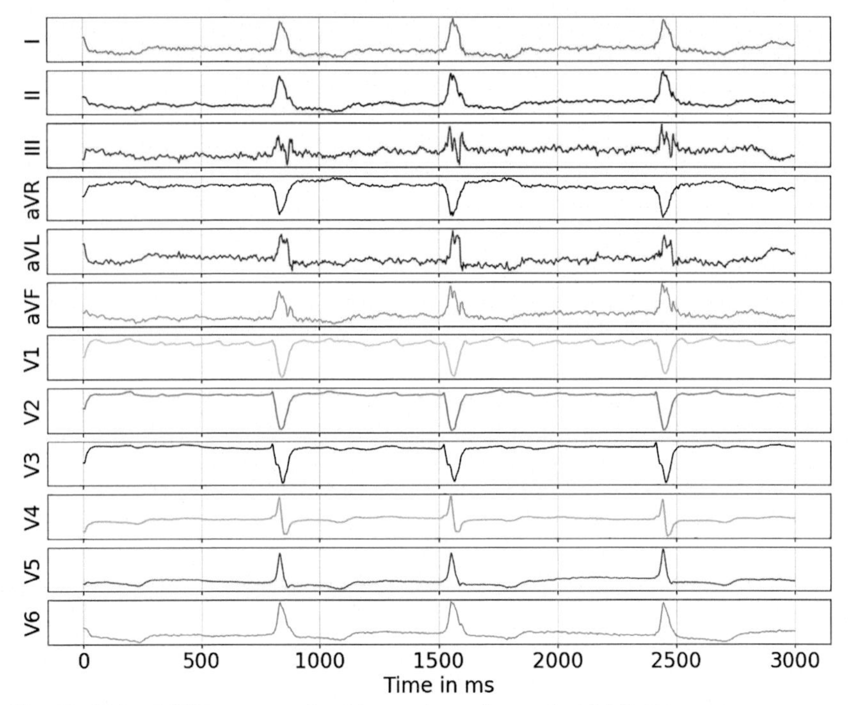

Fig. 10 12-lead ECG annotated as T wave inversion and atrial flutter.

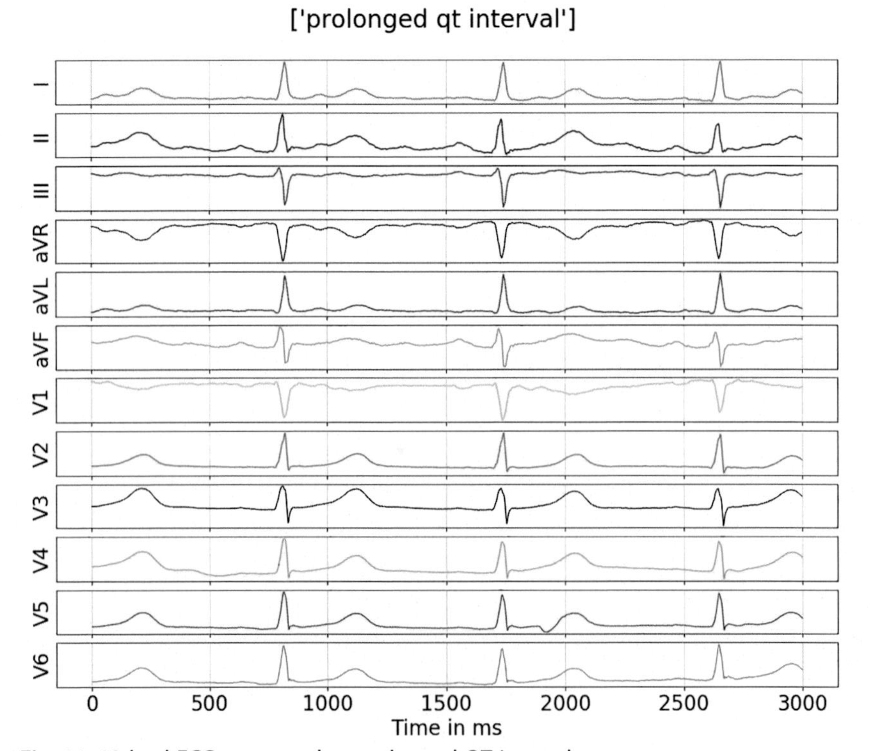

Fig. 11 12-lead ECG annotated as prolonged QT interval.

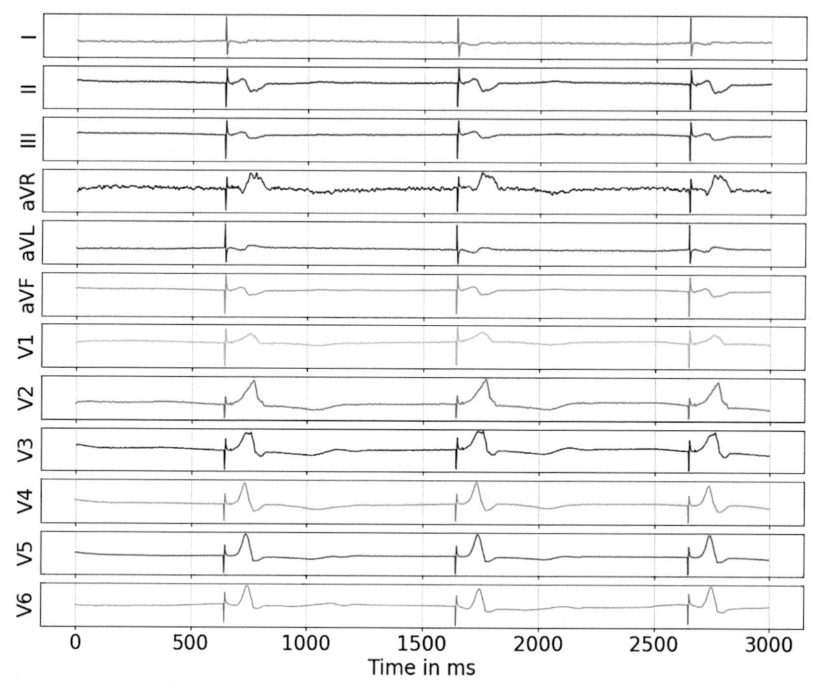

Fig. 12 12-lead ECG annotated as atrial flutter, premature ventricular contractions, and pacing rhythm.

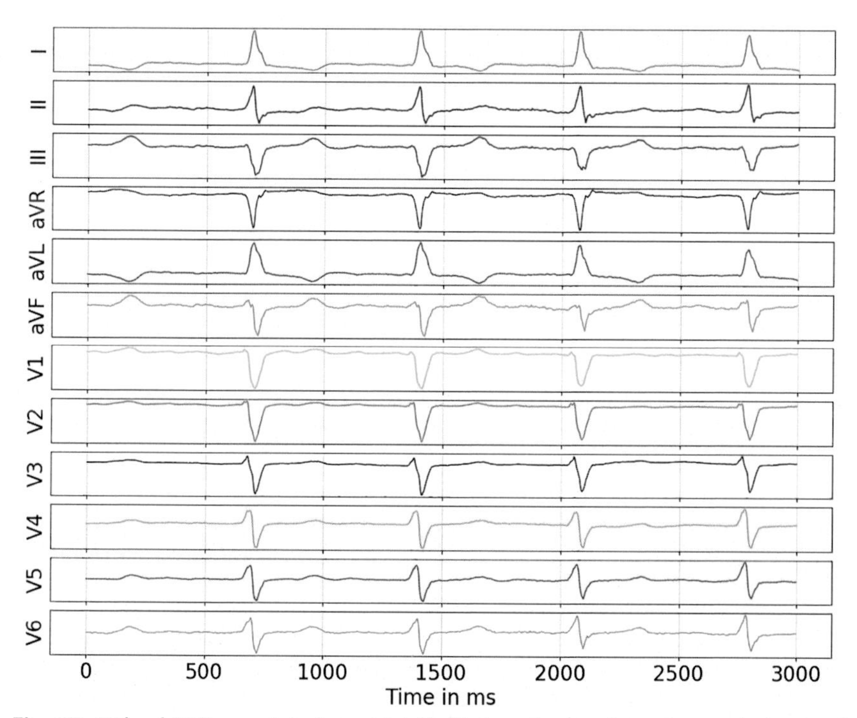

Fig. 13 12-lead ECG annotated as atrial fibrillation, T wave inversion, and abnormal T wave.

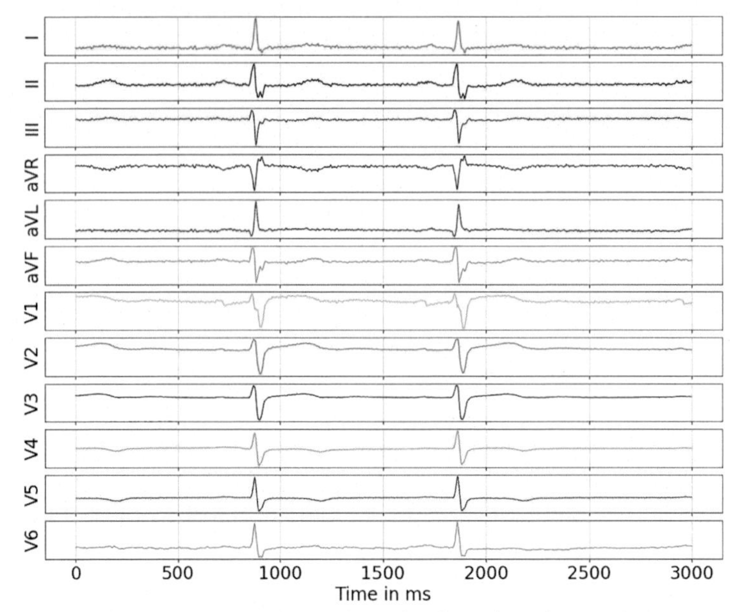

Fig. 14 12-lead ECG annotated as sinus bradycardia, abnormal T wave, sinus arrhythmia, and T wave inversion.

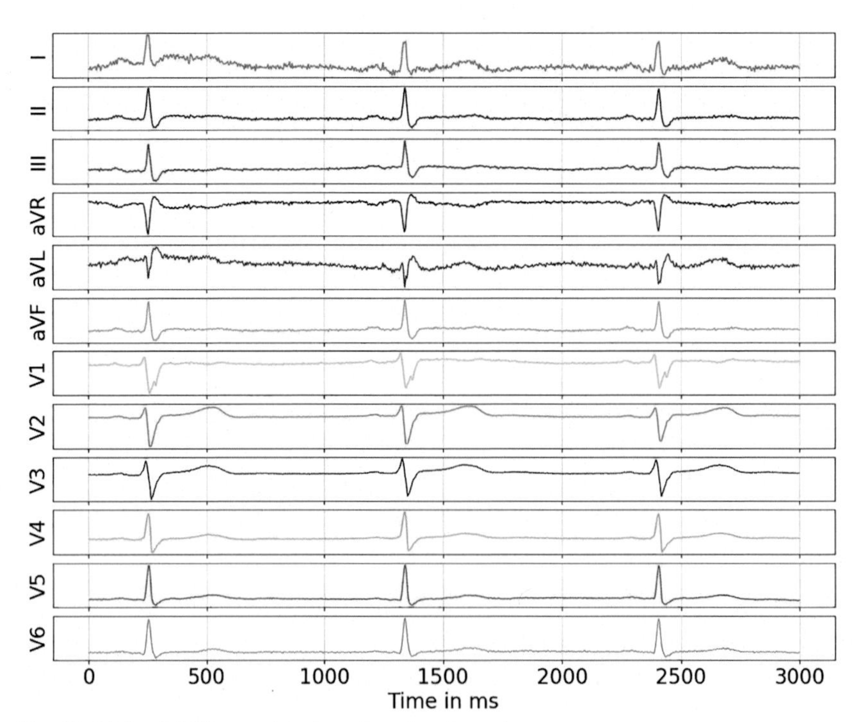

Fig. 15 12-lead ECG annotated as sinus bradycardia.

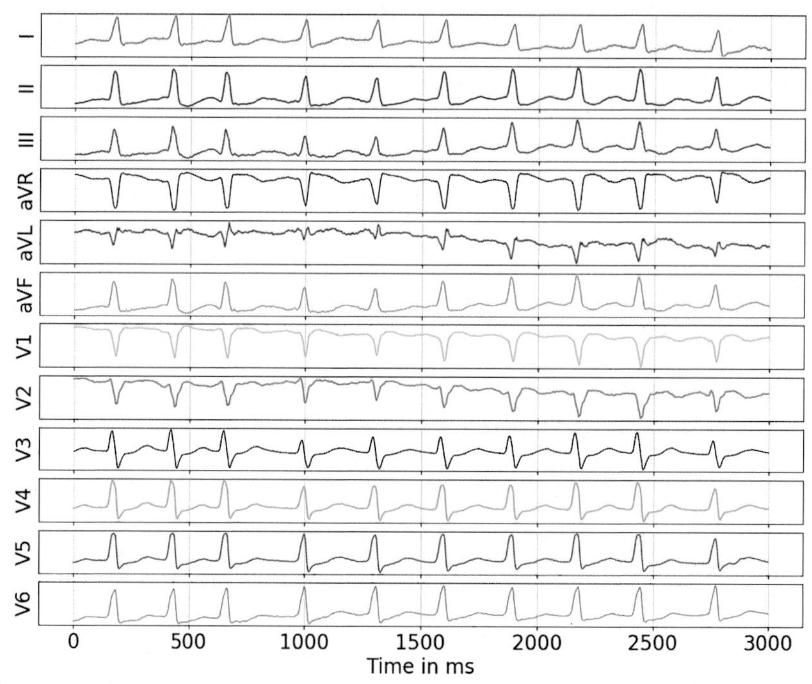

Fig. 16 12-lead ECG annotated as nonspecific intraventricular conduction disorder and atrial fibrillation.

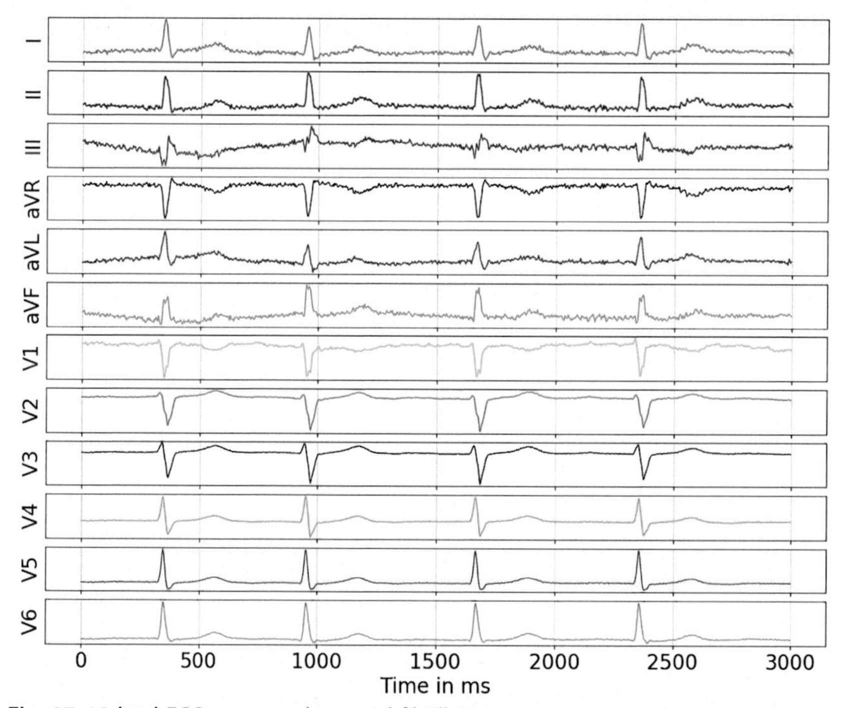

Fig. 17 12-lead ECG annotated as atrial fibrillation.

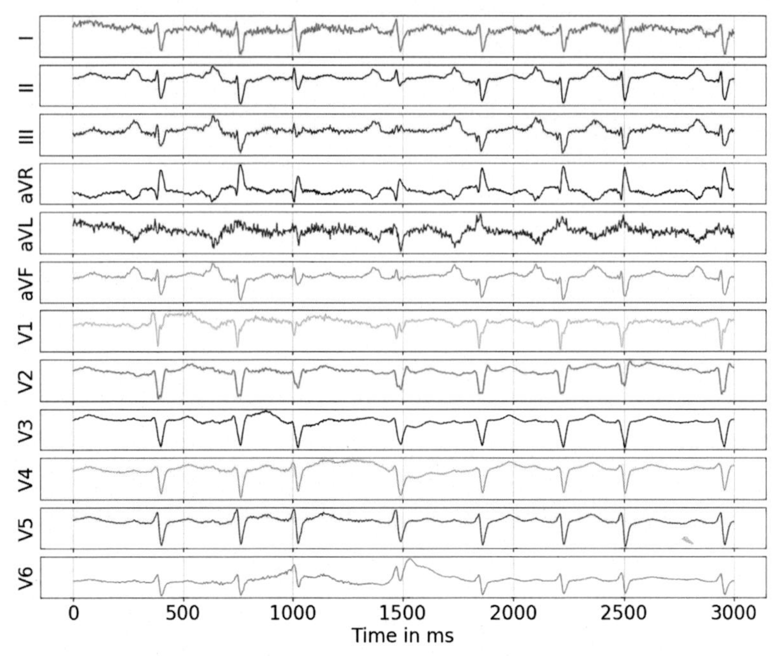

Fig. 18 12-lead ECG annotated as premature atrial contraction, right axis deviation, and sinus tachycardia.

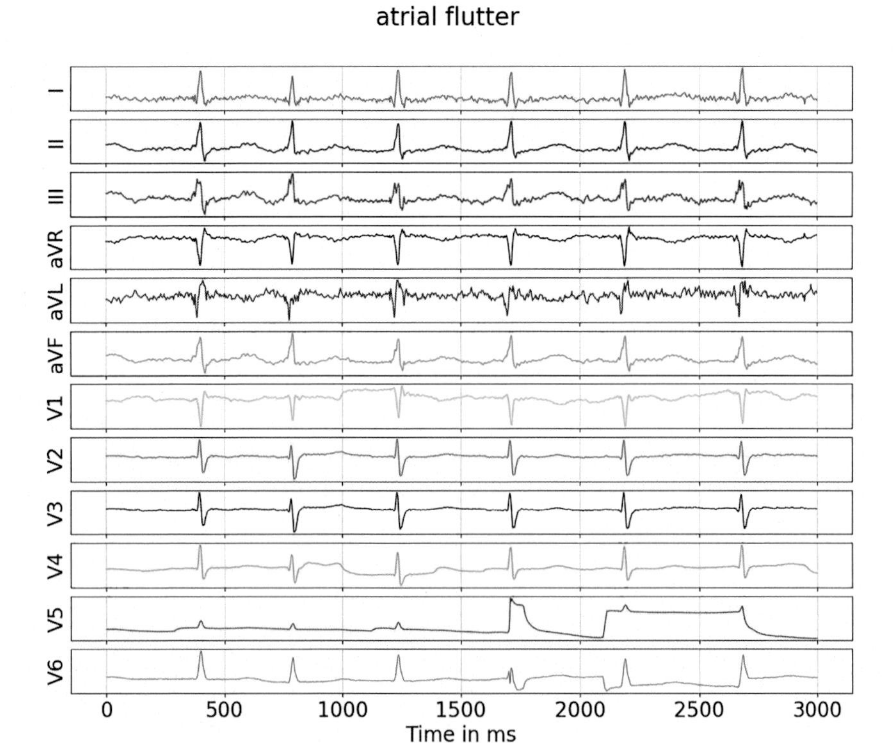

Fig. 19 12-lead ECG annotated as atrial flutter.

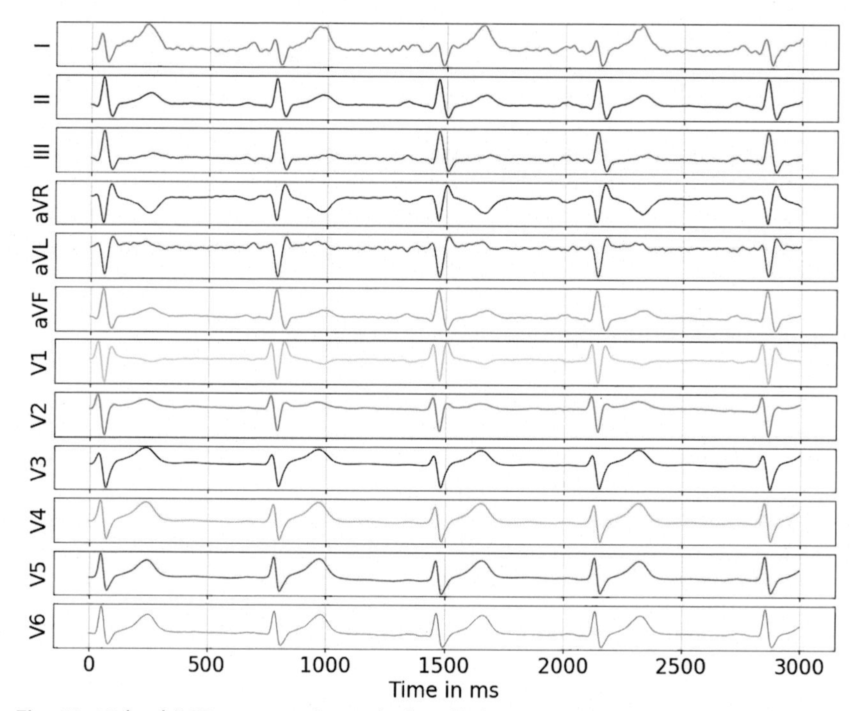

Fig. 20 12-lead ECG annotated as right bundle branch block.

ground truth as much as possible by increasing the interrater agreement of the annotations.

The PhysioNet 2020 Challenge attracted 1395 submissions from 217 teams [45]. The organizers developed a custom metric which incentivizes classifiers able to detect signals with multiple labels (compared to signals having a single label). Post consultation with cardiologists, they also chose to reduce the penalty for misclassification which led to similar treatment or similar risk. Most of the algorithms submitted were different flavors of DNN. The intersubmission analysis further revealed a huge drop in performance on a hidden test set. The problem was identified as overtraining on some of the training dataset and it demonstrates the importance of including multiple data sources to address generalization. In Table 2, we summarize the proposed methodologies of some of the top performing submissions of the PhysioNet 2020. The important observations are following:

Table 2 State-of-the-art methods to detect cardiac disease from 12-lead ECG PhysioNet 2020 dataset

Brief methodology	Result on hidden test data
One-dimensional deep convolutional neural network (CNN) with exponentially dilated causal convolutions is employed. Each causal convolution has multiple causal convolutions, weight normalizations, leaky ReLUs, and residual connections [47].	0.417
A modified ResNet with scatter transform was introduced as the encoder of the data. The encoder module is followed by multihead self attention layer. They also randomly added data augmentation techniques to manage overfitting such as power noise frequencies, Gaussian noise with zero mean and 0.08 standard deviation, sinusoidal drift with random phase, frequency and amplitude resembling baseline drift [48].	0.485
ECG time domain features such as RR interval, RR interval differences, pNN20, pNN50, P wave onset, QT interval ST interval, T wave onset, QRS width were calculated. Also morphological features such as ratio of amplitudes of (R,P), (R,Q), (R,T), and (R/S) peaks were used. An XGBoost classifier algorithm was used to enhance the robustness of the proposed solution [49].	0.233
Authors proposed a 34 layer ResNet architecture having 17 sequential skip connections. One-dimensional convolutions were used to extract features from the 12-lead ECG. The approach also concatenated the demographic features [50].	0.359
The authors proposed modified EfficientNet model. They used random augmentation to increase the robustness of the solution [51].	0.456
In this approach, the authors applied two modeling techniques: A reconstructed phase space Gaussian mixture model (RPS-GMM) and a one-dimensional CNN with 11 layers [52].	0.492
A multiscale shared convolution kernel model was designed to extract deep features of the 12-lead ECG. Squeeze and excitation networks (SE-Net) were added to every path of the model [53].	0.411
A modified ResNet with larger kernel sizes modeling long-term dependencies was proposed in this approach. Also a SE-Net layer was embedded in order to learn the importance of each lead [54].	0.520

Table 2 State-of-the-art methods to detect cardiac disease from 12-lead ECG PhysioNet 2020 dataset—cont'd

Brief methodology	Result on hidden test data
Knowledge of a physician is transformed into a rule-based method using morphological and time-frequency ECG descriptors. In the second phase, a continuous wavelet transform was performed to convert ECGs into an image representing the time-frequency properties (known as a scalogram). Finally, GoogleNet was used to classify. GoogleNet is a 22-layer CNN pretrained classifier created using images of 1000 object categories [55].	0.426
A simple CNN with four convolutional blocks and one fully connected layer was proposed [56].	0.349
In this approach, the authors proposed physiological reasonable feature maps in both amplitude and timing domains. Boosting classifiers were used to generate the predictions. Deep learning methods on the same set of features showed signs of overfitting [57].	0.244
Here the authors used time domain and morphological features involving QRS complex, T wave, and P wave for each of the classes. They also devised rules for each of the target classes. Finally, they used decision trees to classify [58].	0.354
The authors proposed two residual networks with SE-Net blocks to learn the deep features. They also designed a multilabel loss to emphasize wrong prediction by penalizing them. A simple clinical knowledge-driven rule-based bradycardia model was used to achieve robustness [59].	0.514
An 18-layer ResNet was proposed where the residual blocks consist of 11×1 preactivation convolutions with batch normalization and ReLU [60].	0.420
A combination of CNN and LSTM was proposed for feature extraction. Augmentation techniques such as random pad and crop, adding low frequency artifacts, and lead dropout were used to increase generalization [61].	0.437
The authors proposed a hybrid model of deep and wide features. The wide feature module extracts 300 ECG features from Lead II such as time frequency and nonlinear features as well as different morphological features. The deep module consists of an embedding network to extract information from the ECG, a transformer code stack and a multilabel classification. Finally, the wide and deep features were combined in the fully connected layer to produce multilabel classification. [62].	0.533

1. Although a few classical ML models were employed, the bulk of the researchers opted for the deep network approach. There are 2 motivations for this decision: The huge data size of the dataset, and multiclass multilabel consisting of combination of 27 different classes, making it impractical to code specific features for each of the conditions.
2. ResNet and SE-Net were the top performing deep networking models.
3. Interestingly, the top performing model employed a hybrid approach of wide and deep architecture.
4. Deep learning often tends to overfit on the training data. In order to mitigate the problem, most of the algorithms employed a different form of time series augmentation. Please note that these augmentation techniques are fundamentally different from augmentation techniques used in images such as rotation, cropping, bright, contrast, color, etc.
5. Another major challenge was to deal with varying sizes of ECGs (10 s to 30 min). Different techniques such as random selection and zero padding have been employed to make them a similar length.

The PhysioNet 2021 challenge proposed another challenge with the same dataset. The aim of this challenge was to see how accurately one can classify with fewer leads (2-lead, 3-lead, 4-lead, and 6-lead data) [63]. Very interestingly, there was negligible change in performance in terms of the combination of leads. The median change in the challenge performance metrics across different combinations were less than 0.036. In other words, going by these results by a sizable number of participants, a 2-lead analysis is proven to be as accurate as a 12-lead analysis. This is of huge importance in terms of deployment as a 2-lead ECG machine can be ubiquitously worn used 24×7, as opposed to a more invasive 12-lead medical version of it.

5. Conclusion

Every day, automated ECG analysis is becoming more serious and over the past few years, the solution has grown from being a research project into a cloud-based algorithm running on sensory data churned out by mature, medically approved, off-the-shelf devices. In this chapter, we have explained different modalities of noninvasive detection of abnormal heart rhythms using PPG and ECG signals. We have showcased that although PPG is not as informative as ECG, one can find beat-to-beat intervals with acceptable accuracy leading to detection of paroxysmal AF episodes. We have also covered how a more intricate and complex set of cardiac problems can be detected using DNN on single lead or 12-lead ECG signals. We foresee the availability of large ECG datasets as the start of a new biomedical era and as time passes, more geographically and medically diverse data will become available for further research. Future research may focus on utilizing auxiliary information such as subjects' metadata (age, weight, height, personal medical history, family history) during the training phase.

References

[1] B.M. Pluim, A.H. Zwinderman, A. van der Laarse, E.E. van der Wall, The athlete's heart: a meta-analysis of cardiac structure and function, Circulation 101 (3) (2000) 336–344.

[2] C. Antzelevitch, A. Burashnikov, Overview of basic mechanisms of cardiac arrhythmia, Card. Electrophysiol. Clin. 3 (1) (2011) 23–45.

[3] Centers for Disease Control and Prevention, National Center for Health Statistics, Multiple Cause of Death 1999-2017 on CDC WONDER Online Database, October 4, 2019, Released December 2018. Data are from the Multiple Cause of Death Files, 1999-2017, as compiled from data provided by the 57 vital statistics jurisdictions through the Vital Statistics Cooperative Program. Accessed at http://wonder.cdc.gov/mcd-icd10.html.

[4] C.J. Lanzarotti, B. Olshansky, Thromboembolism in chronic atrial flutter: is the risk underestimated? J. Am. Coll. Cardiol. 30 (6) (1997) 1506–1511.

[5] https://www.mayoclinic.org/diseases-conditions/premature-ventricular-contractions/symptoms-causes/syc-20376757. last accessed 13 February 2022.

[6] E.P. Gerstenfeld, T. De Marco, Premature ventricular contractions: not always innocent bystanders, Circulation 140 (8) (2019) 624–626.

[7] A. Baldzizhar, E. Manuylova, R. Marchenko, Y. Kryvalap, M.G. Carey, Ventricular tachycardias: characteristics and management, Crit. Care Nurs. Clin. North Am. 28 (3) (2016) 317–329.

[8] https://www.urmc.rochester.edu/encyclopedia/content.aspx?contenttypeid=134&contentid=234. last accessed 16 February 2022.

[9] H.-K. Ra, J. Ahn, H.J. Yoon, D. Yoon, S.H. Son, J.G. Ko, I am a "smart" watch, smart enough to know the accuracy of my own heart rate sensor, in: Proceedings of the 18th International Workshop on Mobile Computing Systems and Applications, 2017, pp. 49–54.

[10] B. Reeder, A. David, Health at hand: a systematic review of smart watch uses for health and wellness, J. Biomed. Inform. 63 (2016) 269–276.

[11] D. McDuff, S. Gontarek, R.W. Picard, Remote detection of photoplethysmographic systolic and diastolic peaks using a digital camera, IEEE Trans. Biomed. Eng. 61 (12) (2014) 2948–2954.

[12] D. Biswas, N. Simões-Capela, C. Van Hoof, N. Van Helleputte, Heart rate estimation from wrist-worn photoplethysmography: a review, IEEE Sensors J. 19 (16) (2019) 6560–6570.

[13] A.D. Choudhury, A. Visvanathan, R. Banerjee, A. Sinha, A. Pal, C. Bhaumik, A. Kumar, Heartsense: estimating blood pressure and ECG from photoplethysmograph using smart phones, in: Proceedings of the 11th ACM Conference on Embedded Networked Sensor Systems, 2013, pp. 1–2.

[14] Z. Zhang, Z. Pi, B. Liu, TROIKA: a general framework for heart rate monitoring using wrist-type photoplethysmographic signals during intensive physical exercise, IEEE Trans. Biomed. Eng. 62 (2) (2014) 522–531.

[15] Z. Zhang, Photoplethysmography-based heart rate monitoring in physical activities via joint sparse spectrum reconstruction, IEEE Trans. Biomed. Eng. 62 (8) (2015) 1902–1910.
[16] M.N.K. Lakshminarasimha, P.C. Madhusudana, P. Suresha, V. Periyasamy, P.K. Ghosh, Multiple spectral peak tracking for heart rate monitoring from photoplethysmography signal during intensive physical exercise, IEEE Signal Process Lett. 22 (12) (2015) 2391–2395.
[17] A. Temko, Accurate heart rate monitoring during physical exercises using PPG, IEEE Trans. Biomed. Eng. 64 (9) (2017) 2016–2024.
[18] S. Salehizadeh, D. Dao, J. Bolkhovsky, C. Cho, Y. Mendelson, K.H. Chon, A novel time-varying spectral filtering algorithm for reconstruction of motion artifact corrupted heart rate signals during intense physical activities using a wearable photoplethysmogram sensor, Sensors 16 (1) (2016) 10.
[19] H. Chung, H. Lee, J. Lee, Finite state machine framework for instantaneous heart rate validation using wearable photoplethysmography during intensive exercise, IEEE J. Biomed. Health Inform. 23 (4) (2018) 1595–1606.
[20] M.B. Mashhadi, E. Asadi, M. Eskandari, S. Kiani, F. Marvasti, Heart rate tracking using wrist-type photoplethysmographic (PPG) signals during physical exercise with simultaneous accelerometry, IEEE Signal Process Lett. 23 (2) (2015) 227–231.
[21] E. Khan, F. Al Hossain, S.Z. Uddin, S.K. Alam, M.K. Hasan, A robust heart rate monitoring scheme using photoplethysmographic signals corrupted by intense motion artifacts, IEEE Trans. Biomed. Eng. 63 (3) (2015) 550–562.
[22] D. Zhao, S. Yu, S. Wan, F. Wang, SFST: a robust framework for heart rate monitoring from photoplethysmography signals during physical activities, Biomed. Signal Process. Control 33 (2017) 316–324.
[23] D. Biswas, L. Everson, M. Liu, M. Panwar, B.-E. Verhoef, S. Patki, C.H. Kim, et al., CorNET: deep learning framework for PPG-based heart rate estimation and biometric identification in ambulant environment, IEEE Trans. Biomed. Circuits Syst. 13 (2) (2019) 282–291.
[24] T. Bhattacharjee, A.D. Choudhury, A. Pal, Robust beat-to-beat interval from wearable PPG using RLS and SSA, in: 2019 41st Annual International Conference of the IEEE Engineering in Medicine and Biology Society (EMBC), IEEE, 2019, pp. 4946–4952.
[25] A. Mukherjee, A.D. Choudhury, S. Datta, C. Puri, R. Banerjee, R. Singh, A. Ukil, S. Bandyopadhyay, A. Pal, S. Khandelwal, Detection of atrial fibrillation and other abnormal rhythms from ECG using a multi-layer classifier architecture, Physiol. Meas. 40 (5) (2019) 054006.
[26] J. Pan, W.J. Tompkins, A real-time QRS detection algorithm, IEEE Trans. Biomed. Eng. 3 (1985) 230–236.
[27] A.E.W. Johnson, J. Behar, F. Andreotti, G.D. Clifford, J. Oster, Multimodal heart beat detection using signal quality indices, Physiol. Meas. 36 (8) (2015) 1665.
[28] J. Park, S. Lee, M. Jeon, Atrial fibrillation detection by heart rate variability in Poincare plot, Biomed. Eng. Online 8 (1) (2009) 1–12.
[29] S. Sarkar, D. Ritscher, R. Mehra, A detector for a chronic implantable atrial tachyarrhythmia monitor, IEEE Trans. Biomed. Eng. 55 (3) (2008) 1219–1224.
[30] G. Moody, A new method for detecting atrial fibrillation using RR intervals, Comput. Cardiol. (1983) 227–230.
[31] S. Cerutti, L.T. Mainardi, A. Porta, A.M. Bianchi, Analysis of the dynamics of RR interval series for the detection of atrial fibrillation episodes, Comput. Cardiol. (1997) 77–80. IEEE.
[32] K. Tateno, L. Glass, Automatic detection of atrial fibrillation using the coefficient of variation and density histograms of RR and ΔRR intervals, Med. Biol. Eng. Comput. 39 (6) (2001) 664–671.

[33] S. Dash, K.H. Chon, S. Lu, E.A. Raeder, Automatic real time detection of atrial fibrillation, Ann. Biomed. Eng. 37 (9) (2009) 1701–1709.
[34] S. Babaeizadeh, R.E. Gregg, E.D. Helfenbein, J.M. Lindauer, S.H. Zhou, Improvements in atrial fibrillation detection for real-time monitoring, J. Electrocardiol. 42 (6) (2009) 522–526.
[35] J. Oster, G.D. Clifford, Impact of the presence of noise on RR interval-based atrial fibrillation detection, J. Electrocardiol. 48 (6) (2015) 947–951.
[36] R. Colloca, A.E.W. Johnson, L. Mainardi, G.D. Clifford, A support vector machine approach for reliable detection of atrial fibrillation events, in: Computing in Cardiology 2013, IEEE, 2013, pp. 1047–1050.
[37] A.L. Goldberger, L.A.N. Amaral, L. Glass, J.M. Hausdorff, P.C. Ivanov, R.G. Mark, J. E. Mietus, G.B. Moody, C.-K. Peng, H.E. Stanley, PhysioBank, PhysioToolkit, and PhysioNet: components of a new research resource for complex physiologic signals, Circulation 101 (23) (2000) e215–e220.
[38] M. Zabihi, A.B. Rad, A.K. Katsaggelos, S. Kiranyaz, S. Narkilahti, M. Gabbouj, Detection of atrial fibrillation in ECG hand-held devices using a random forest classifier, in: 2017 Computing in Cardiology (CinC), IEEE, 2017, pp. 1–4.
[39] S. Hong, M. Wu, Y. Zhou, Q. Wang, J. Shang, H. Li, J. Xie, ENCASE: an ENsemble ClASsifiEr for ECG classification using expert features and deep neural networks, in: 2017 Computing in cardiology (CinC), IEEE, 2017, pp. 1–4.
[40] A. Sagie, M.G. Larson, R.J. Goldberg, J.R. Bengtson, D. Levy, An improved method for adjusting the QT interval for heart rate (the Framingham Heart Study), Am. J. Cardiol. 70 (7) (1992) 797–801.
[41] V. Kalidas, L.S. Tamil, Cardiac arrhythmia classification using multi-modal signal analysis, Physiol. Meas. 37 (8) (2016) 1253.
[42] S. Yazdani, J.-M. Vesin, Adaptive mathematical morphology for QRS fiducial points detection in the ECG, in: Computing in Cardiology 2014, IEEE, 2014, pp. 725–728.
[43] L.M. Eerikäinen, J. Vanschoren, M.J. Rooijakkers, R. Vullings, R.M. Aarts, Decreasing the false alarm rate of arrhythmias in intensive care using a machine learning approach, in: 2015 Computing in Cardiology Conference (CinC), IEEE, 2015, pp. 293–296.
[44] G.D. Clifford, C. Liu, B. Moody, H.L. Li-wei, I. Silva, Q. Li, A.E. Johnson, R.G. Mark, AF classification from a short single lead ECG recording: the PhysioNet/computing in cardiology challenge 2017, in: 2017 Computing in Cardiology (CinC), IEEE, 2017, pp. 1–4.
[45] A.E.A. Perez, A. Gu, A.J. Shah, C. Robichaux, A.-K.I. Wong, C. Liu, F. Liu, et al., Classification of 12-lead ECGs: the physionet/computing in cardiology challenge 2020, Physiol. Meas. 41 (12) (2020) 124003.
[46] A.Y. Hannun, P. Rajpurkar, M. Haghpanahi, G.H. Tison, C. Bourn, M.P. Turakhia, Y.N. Andrew, Cardiologist-level arrhythmia detection and classification in ambulatory electrocardiograms using a deep neural network, Nat. Med. 25 (1) (2019) 65–69.
[47] M.N. Bos, R.R. van de Leur, J.F. Vranken, D.K. Gupta, P. van der Harst, P.A. Doevendans, R. van Es, Automated comprehensive interpretation of 12-lead electrocardiograms using pre-trained exponentially dilated causal convolutional neural networks, in: 2020 Computing in Cardiology, IEEE, 2020, pp. 1–4.
[48] M.P. Oppelt, M. Riehl, F.P. Kemeth, J. Steffan, Combining scatter transform and deep neural networks for multilabel electrocardiogram signal classification, in: 2020 Computing in Cardiology, IEEE, 2020, pp. 1–4.
[49] D.U. Uguz, F. Berief, S. Leonhardt, C.H. Antink, Classification of 12-lead ECGs using gradient boosting on features acquired with domain-specific and domain-agnostic methods, in: 2020 Computing in Cardiology, IEEE, 2020, pp. 1–4.

[50] W. Jia, X. Xu, X. Xu, Y. Sun, X. Liu, Automatic detection and classification of 12-lead ECGs using a deep neural network, in: 2020 Computing in Cardiology, IEEE, 2020, pp. 1–4.
[51] N. Nonaka, J. Seita, Electrocardiogram classification by modified EfficientNet with data augmentation, in: 2020 Computing in Cardiology, IEEE, 2020, pp. 1–4.
[52] D. Kaftan, R.J. Povinelli, A deep neural network and reconstructed phase space approach to classifying 12-lead ECGs, in: 2020 Computing in Cardiology, IEEE, 2020, pp. 1–4.
[53] J. Chen, T. Chen, B. Xiao, X. Bi, Y. Wang, H. Duan, W. Li, J. Zhang, M. Xu, SE-ECGNet: multi-scale SE-Net for multi-lead ECG data, in: 2020 Computing in Cardiology, IEEE, 2020, pp. 1–4.
[54] Z. Zhao, H. Fang, S.D. Relton, R. Yan, Y. Liu, Z. Li, J. Qin, D.C. Wong, Adaptive lead weighted resnet trained with different duration signals for classifying 12-lead ECGs, in: 2020 Computing in Cardiology, IEEE, 2020, pp. 1–4.
[55] G. Bortolan, I. Christov, I. Simova, Rule-based method and deep learning networks for automatic classification of ECG, in: 2020 Computing in Cardiology, IEEE, 2020, pp. 1–4.
[56] M. Soliński, M. Lepek, A. Pater, K. Muter, P. Wiszniewski, D. Kokosińska, J. Salamon, Z. Puzio, 12-lead ECG arrythmia classification using convolutional neural network for mutually non-exclusive classes, in: 2020 Computing in Cardiology, IEEE, 2020, pp. 1–4.
[57] P.-Y. Hsu, P.-H. Hsu, T.-H. Lee, H.-L. Liu, Multi-label arrhythmia classification from 12-lead electrocardiograms, in: 2020 Computing in Cardiology, IEEE, 2020, pp. 1–4.
[58] R. Smisek, A. Nemcova, L. Marsanova, L. Smital, M. Vitek, J. Kozumplik, Cardiac pathologies detection and classification in 12-lead ECG, in: 2020 Computing in Cardiology, IEEE, 2020, pp. 1–4.
[59] Z. Zhu, H. Wang, T. Zhao, Y. Guo, Z. Xu, Z. Liu, S. Liu, X. Lan, X. Sun, M. Feng, Classification of cardiac abnormalities from ECG signals using SE-ResNet, in: 2020 Computing in Cardiology, IEEE, 2020, pp. 1–4.
[60] S. Min, H.-S. Choi, H. Han, M. Seo, J.-K. Kim, J. Park, S. Jung, O. Il-Young, B. Lee, S. Yoon, Bag of tricks for electrocardiogram classification with deep neural networks, in: 2020 Computing in Cardiology, IEEE, 2020, pp. 1–4.
[61] H. Hasani, A. Bitarafan, M.S. Baghshah, Classification of 12-lead ECG signals with adversarial multi-source domain generalization, in: 2020 Computing in Cardiology, IEEE, 2020, pp. 1–4.
[62] A. Natarajan, Y. Chang, S. Mariani, A. Rahman, G. Boverman, S. Vij, J. Rubin, A wide and deep transformer neural network for 12-lead ECG classification, in: 2020 Computing in Cardiology, IEEE, 2020, pp. 1–4.
[63] M.A. Reyna, N. Sadr, E.A. Perez Alday, A. Gu, A.J. Shah, C. Robichaux, A.B. Rad, et al., Will two do? Varying dimensions in electrocardiography: the PhysioNet/Computing in Cardiology Challenge 2021, in: 2021 Computing in Cardiology (CinC), vol. 48, IEEE, 2021, pp. 1–4.

Further reading

R. Krishnan, B. Natarajan, S. Warren, Two-stage approach for detection and reduction of motion artifacts in photoplethysmographic data, IEEE Trans. Biomed. Eng. 57 (8) (2010) 1867–1876.

CHAPTER 5

Heart blockage

1. Introduction

The heart acts like a pump that continuously supplies oxygenated blood to the organs. A normal heart beats 60–100 times every minute. Coronary arteries are responsible for supplying oxygen-rich blood to the heart muscles to carry out their normal functions. The coronary arteries are divided into tiny multiple branches to reach the heart muscles. The human heart has two main coronary arteries responsible for blood flow, the left main coronary arteries and the right coronary arteries. The left main coronary artery supplies blood to the left side of the muscles including the left atrium and the left ventricle. The right coronary artery is responsible for supplying blood to the right part of the heart including right atrium, right ventricle, and the sinoatrial (SA) and atrioventricular (AV) nodes. The main coronary arteries are branched into smaller arteries to supply blood to the other parts of the heart. In order to ensure normal blood circulation through coronary arteries, the arterial paths need to be absolutely clean. However, there are situations when the arterial paths are blocked due to blood clots or deposition of fatty materials such as cholesterol, commonly known as plaque, over time. Depending upon the amount of blockage in the coronary arteries, the consequences can vary from mild to severe cardiac conditions. A small degree of blockage may not cause any symptoms at all, whereas a high degree of heart blockage may cause the heart to have difficulty in pumping enough blood to the circulatory system. As a result, the heart muscles, the brain, and other organs may be deprived of a sufficient amount of oxygen-rich blood. This may end up causing a stroke or severe cardiac arrest and death. Statistics reveal that around 800,000 people suffer from heart attacks every year.

The most common early sign of a heart blockage is angina or chest pain, which may also spread to the back, arms, or jaw. Apart from this, dizziness, shortness of breath, nausea, and rapid irregular heartbeats are considered other important symptoms for a potential cardiac blockage. However, the major problem is that heart blockages are often asymptomatic in the early

New Frontiers of Cardiovascular Screening using Unobtrusive Sensors, AI, and IoT
https://doi.org/10.1016/B978-0-12-824499-9.00005-2

stages. Hence, many patients delay an early diagnosis and often seek medical advice only at an advanced stage of a critical disease when the symptoms strongly appear. This not only makes the treatment costly and complicated, but also increases the chances of mortality. Consequently, doctors often suggest an early diagnosis of heart blockage-related issues. In general, the coronary arteries are flexible enough to handle the blood flow through. However, the natural elasticity of the arteries decays with age. Hence, the small arteries may not handle normal blood flow. Moreover, the artery diameter may reduce due to deposition of plaque over time. This is why the elderly are in general more vulnerable to developing a heart block. However, heart blockage is widely considered as a lifestyle disease. Unfortunately, in recent times, a large number of young adults at their early thirties are also found to be at high risk of developing cardiac diseases due to a sedentary lifestyle, lack of regular physical activities, and an unhealthy diet. Some people are born with a heart block. However, older people with a past family history of heart disease or smoking are most at risk. Conditions such as cardiomyopathy, coronary thrombosis, myocarditis, endocarditis, or inflammation of the heart valves or scar tissue in the heart due to prior heart surgery, enhance the risk factor for a heart block. It is to be noted that a severe heart blockage cannot be cured by medication, hence it requires a cardiac surgery such as a coronary angioplasty or a coronary bypass.

Fig. 1 shows a labeled cross-sectional view of the human heart with sample diagrams of the coronary arteries in three different stages of heart blockage. Under normal conditions, there should not be any build up inside the artery (lumen) so that the blood flow is unrestricted (the left-most figure). However, with aging and lifestyle decisions, fatty materials or plaque starts building up in the lumen which partially restricts the blood flow (the middle figure). If the condition is not detected at this stage and remains untreated, the plaque will grow further and completely block the blood flow eventually (the right-most figure). This may lead to a severe cardiac arrest.

According to doctors, there are three types of heart block.

A first-degree heart block can be considered the primary stage of heart blockage. It typically involves minor heartbeat disruptions, such as skipped beats or mild irregular heartbeats and can also be asymptomatic. It is considered to be the least serious type of heart block, and it does not generally require treatment. In this stage, a doctor typically asks the patient for regular checkups and prescribes medications for the thinning of the blood and controlling erratic heart beats.

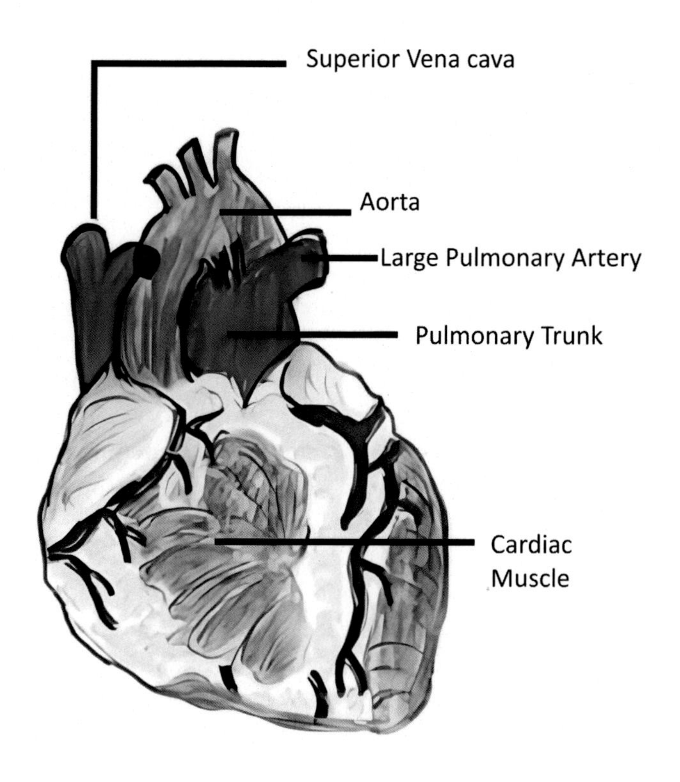

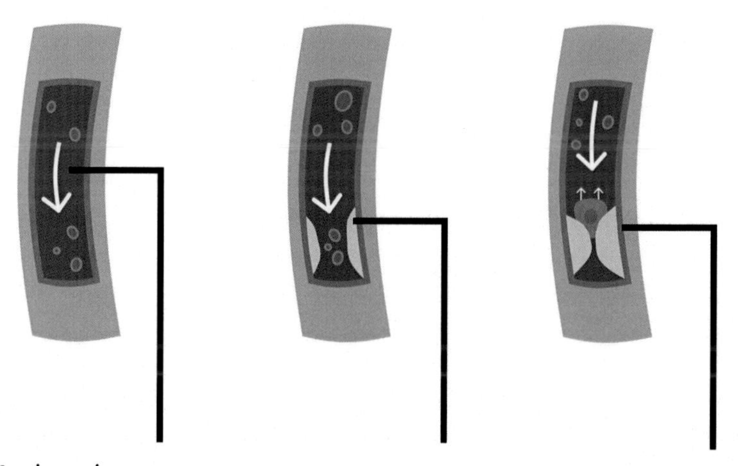

Fig. 1 Condition of coronary artery at different stages of heart blockage.

A second-degree heart block is more serious, and it occurs when some electrical signals do not reach the right portion of the heart, causing dropped or skipped beats. The patient may need a pacemaker.

Third-degree or a complete heart blockage is more severe, and it occurs when the regular electrical signals generated in the heart cannot travel between the upper and lower chambers due to arterial blockages. It is more common in patients with heart disease. Because there is a serious risk of heart attack, doctors often suggest implanting a pacemaker in this stage.

Coronary artery disease (CAD) is a common type of condition due to blockages in the coronary arteries. The human heart has four main coronary arteries responsible for blood flow. These are the right coronary artery, the left coronary artery, the left anterior descending artery, and the left circumflex artery. CAD is formed due to the gradual deposition of cholesterol and other types of fatty materials on the inner walls of the coronary arteries. This condition is called atherosclerosis. As a result, the coronary walls are stiffened, and the artery path is blocked, limiting the blood flow to heart muscles. Hence, the heart muscles do not receive sufficient oxygen and other nutrients to carry out their normal functions. This condition is called ischemia, which causes angina or chest pain and enhances the risk of a stroke or a heart attack. Available statistics show that approximately 360,000 people died of CAD in the United States in 2019. About 18.2 million people, aged above 20, develop CAD every year and about 2 in every 10 deaths from CAD every year happen in adults who are less than 65 years of age [1].

Angina or chest pain is considered a primary symptom for a heart block. Doctors typically check the intensity of the chest pain along with the patient's age, family history, and lifestyle when suspecting heart disease, followed by several clinical tests for diagnosis of a heart block. The diagnosis typically starts with a standard 12-lead electrocardiogram (ECG) test. An early indication of whether the left or right bundle branch is affected can be determined from the abnormalities in the ECG trace and the corresponding lead location. However, such abnormalities in the ECG trace are often intermittent, and hence not always guaranteed to be captured in a single ECG test. Hence, the doctor might suggest a Holter tape. This is a portable ECG device that the patient wears for 1 or 2 days which continuously records all the electrical activities of the heart even when the patient is undertaking their regular day-to-day activities. This is a good test for identifying intermittent arrhythmias and other episodic abnormal heart rhythms. An exercise stress test or treadmill test can also indicate the presence of heart blockage. Doctors may also suggest an echocardiogram test which is an ultrasonic scan of the heart or a computed tomography (CT) scan for revealing the condition of the heart valves and heart muscles.

The gold standard diagnosis for arterial diagnosis is the coronary angiogram. This is an invasive clinical test that not only lets the cardiologists know the extent of coronary artery blockage, but also diagnoses heart conditions such as an aneurysm or other heart defects. A coronary angiogram involves minor surgery requiring a hospital stay of 1–2 days. During the procedure, doctors insert a small catheter tube into the blood vessels through the wrist or groin after following the necessary local anesthetic procedure. Once the catheter reaches the heart, it injects a small amount of X-ray sensitive dye. Subsequently, a series of dynamic X-rays are taken of the blood vessels as the dye moves through them, revealing the condition of the vessels.

An early, efficient yet noninvasive, detection of heart blockage has become an important requirement in medicine. The recently published research works suggest that an accurate prediction or estimation of cardiac blockage is possible thanks to the advances in digital technologies including artificial intelligence (AI) and machine learning (ML). In this chapter, we will discuss the markers that can be found in noninvasive biomarkers such as ECG and phonocardiogram (PCG) that are known to be correlated with heart blockage, more specifically CAD. We will also discuss how these markers can be used to train a ML algorithm for classification of normal people and CAD patients. We will also discuss how deep learning can be used for identifying CAD.

2. Correlation of heart blockage with ECG, PPG, and PCG

A coronary angiogram, the gold standard clinical test, is invasive and requires admission to hospital for a short stay. It also carries associated risks in terms of anesthesia, infection, and possible blood loss. Such techniques cannot be used as mass screening tools. Thus, a noninvasive early screening for heart disease or CAD using minimum infrastructure is required. Such solutions can be very useful for affordable and quick mass screening in the underdeveloped and developing nations, where the doctor-to-patient ratio is less than desired, and in developed nations for continuous monitoring of the elderly population facing mobility issues to visit clinics.

Fortunately, recent research works published in major scientific forums show the presence of discriminating markers in various cardiovascular signals even at the onset of cardiac diseases.

12-lead ECG recordings are considered an important source for identifying early signs of cardiac abnormalities. Anomalous heart rhythms due to conditions such as ischemia can be identified from an ECG. Additional issues including signs of a heart attack, thickened heart muscle, signs of a previous heart attack, or long-term damage from an electrolyte imbalance can also be identified and approximately localized from the ECG leads. Myocardial repolarization is mainly affected due to ischemia. This may cause morphological changes in the ECG pattern in terms of ST segment depression, ST segment elevation, T wave inversion, or flattening of T waves. Although these morphological changes are not the definitive markers for ischemia, they can be helpful in generating alerts for a detailed diagnosis. On the other hand, an abnormally high Q wave may suggest a possible cardiac infarction.

Fig. 2 shows a single-lead ECG waveform of a 55-year-old patient with cardiac blockage. The change in morphology in ST segment is clearly visible in the image. Hemodialysis (HD) patients are known to have a high rate of cardiac morbidity and mortality. In this condition, the coronary artery oxygen delivery is reduced while the myocardial oxygen demand is substantially increased. As a result, both symptomatic and silent ischemic heart disease may frequently occur [2], and the authors have successfully evaluated the usefulness of a significant ST depression induced by HD for the diagnosis

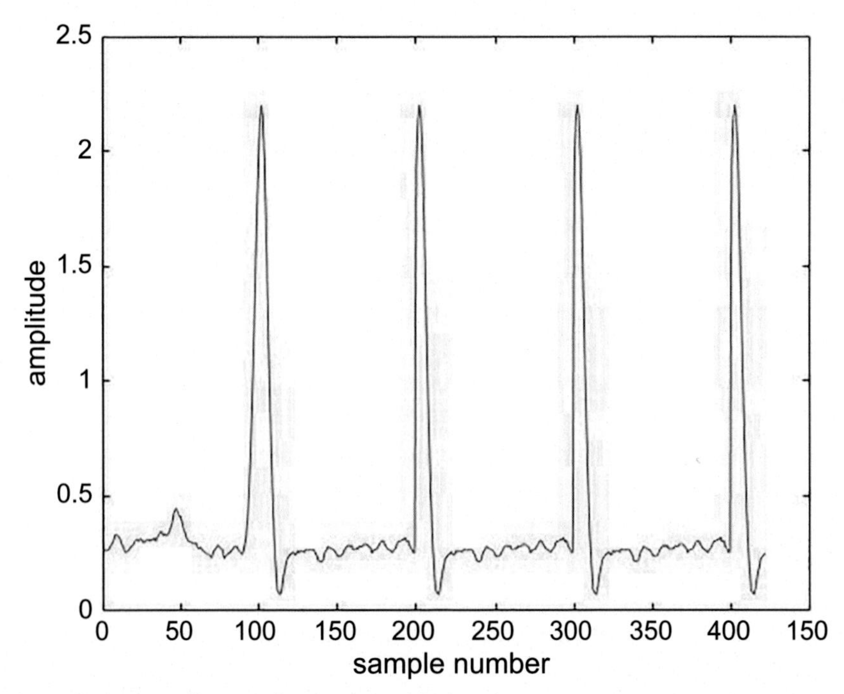

Fig. 2 ECG trace of patient having heart blockage.

of CAD and nominated it as the predictor of subsequent cardiac events in HD patients.

Heart rate variability (HRV) is also considered a predictive factor for a possible cardiac event. Several research articles have shown that an abnormally reduced HRV can be a potential risk factor of cardiac events. CAD patients are known to have a reduced HRV. In Ref. [3], the authors have proposed a number of numerical features to quantify the extent of HRV that can be used to classify CAD patients. In Ref. [4], the authors have proposed a decision-support system which separates two classes (CAD and non-CAD patients) using HRV features obtained from ECGs. Such solutions aim to assist cardiologists during the diagnosis to create a base for a new computer-aided diagnosis system. Although an ECG is considered to be the most efficient source for extracting HRV, 12-lead ECG recording is not the most unobtrusive way. A photoplethysmography (PPG) signal can be considered a cheap yet reliable alternative resource for measuring HRV without an ECG. The peak-to-peak interval distances of PPG highly correlates to the RR interval distances obtained from an ECG. Moreover, several other important parameters can be extracted from PPG for identifying cardiac risk. A lower systolic amplitude of a PPG pulse is an indication of a decreased blood volume pulsations or a decreased venous blood volume. Some of the important PPG features for identifying heart diseases are:

- Crest time: This is the time difference between a PPG trough point and its immediate next peak. This is considered an important parameter for identifying cardiac risk.
- Stiffness index: The time delay between the systolic and diastolic peaks in a PPG pulse is related to the transit time of pressure waves from the root of the subclavian artery to the apparent site of reflection and back to the subclavian artery. The stiffness index (SI) is the ratio of the height of a person to the time delay between the systolic and the diastolic peak.

Although existing literature suggests that a PPG can be considered an important resource for identifying CAD, clinical interpretation of PPG-based CAD markers is not yet fully known. Moreover, narrow band signals such as an ECG are not expected to carry significant information for the accurate detection of CAD.

The heart sound is widely considered an important marker for detecting abnormal behavior of the heart. Doctors heavily rely upon listening to the heart sound for initial assessment of the heart. However, abnormalities in heart sound due to CAD may occur in the higher spectrums, which may not be detectable to the human ear. The advances in digital signal processing and ML techniques can ensure an effective analysis of digitally recorded heart sound for detection of CAD or other types of valvular diseases.

A PCG or the digitally recorded heart sound is thus another important signal for identifying heart diseases. A complete cardiac cycle in a PCG signal has four major components: S1, systole, S2, and diastole. S1 is termed the first heart sound that occurs just before systole, and S2 is termed the second heart sound that occurs just before diastole. Additionally, there are other heart sounds which are related to some types of abnormalities. In general, the PCG signal recorded from a healthy person contains two prominent heart sounds, the S1 and the S2. However, there are two additional heart sounds, S3 and S4, that point to various pathological conditions [5]. The presence of an S3 or S4 heart sound is a potential indication of cardiac disease. Moreover, the diastolic region of the heart sound of patient having a heart blockage often contains high frequency components at a frequency region above 100 Hz. Although they are faint and often not captured by the human auditory system, they can be detected using digital signal processing or advanced artificial intelligence (AI) analytical tools on high quality digital PCG recordings. A number of conditions such as valvular dysfunctions or arterial blockages can be identified from the high frequency components present in the PCG signals.

Several studies suggest that the diastolic portion in the PCG signal contains an important marker for identifying CAD. In Refs. [6,7], the authors modeled PCG diastolic portion in PCG using autoregressive (AR) models and showed that the model coefficients are different for CAD patients compared to non-CAD patients.

3. AI-based detection of chronic ischemic heart disease

So far, we have discussed heart blockage, its type, its consequences, and clinical diagnosis techniques. We have also briefly pointed out that there can be discriminatory markers found in various biomedical signals for early but noninvasive detection of a cardiac event. However, in practice, accurate detection of heart disease from such noninvasive sources is not that easy. The potential biomarkers available in various sensor signals as mentioned in the previous section are not fully clinically accepted as their presence varies from patient to patient and are not consistent at various stages of the diseases. Second, there are no known definite biomarkers for accurately identifying CAD or any type of heart blockage. All the markers discussed are often intermittent in nature, i.e., the markers may not be present in the biosignals all of the time. Further, there is no guarantee that the markers would be present at the onset of the disease when the disease condition is mild. That is why it is practically impossible for a doctor to interpret a disease such as CAD, atherosclerosis, ischemia, or other types of heart blockages from a large volume of data. Here, AI can play an important role as it is very good at continuous monitoring of data for possible identification of a disease. Off-the-shelf devices and solutions are already available for commercial use that are capable of identifying cardiac events. The Apple Watch Series 6 can be considered a great example of the implementation of AI in digital healthcare that one can use in everyday life. However, there are complex AI applications as well running on commercially successful medical-grade equipment in hospital intensive care units (ICUs) for assisting doctors in their critical decision making.

In previous chapters, we have discussed in detail how AI and ML can be used in automatic diagnosis in cardiology to classify or predict a certain disease. Any ML system typically involves some basic steps, i.e., extraction of features from raw data. In biomedical applications, the extraction of features from medical sensors may involve some preprocessing. Depending upon your application, the input data can be of various types. For example, in the case of ECG or digital heart sound data, the input can be time-series data. On the other hand, for analysis of echocardiograms, the inputs can be a series of digital images. In classical ML approaches, you first need to extract the set of parameters, termed features, that are expected to be discriminating for various target classes. In deep learning, you feed your deep learning engine with a large amount of data along with target labels, hoping that the machine will automatically learn the features during training. In the past few chapters, we have talked about noninvasive biomedical signals such as ECG, PCG, and PPG and how we can derive meaningful features from them, that represent the nature of the signals. In this section, we have discussed various biomarkers which can be considered to be features for classifying CAD and non-CAD people. In the following section, we will detail how we can predict the presence or absence of cardiac disease based on this signal. The classification techniques can be broadly divided into two categories: ML approaches and deep learning approaches.

3.1 Machine learning approaches

Machine learning approaches are older, simpler, yet very accurate, and explainable and hence are acceptable for deployment. In the previous section we discussed several biomarkers which are clinically known to be discriminative for CAD patients compared to non-CAD subjects. Those markers can be used as features to train ML algorithms such as support vector machine (SVM) and decision trees for classification of CAD. Numerical features that can be extracted from PPG, PCG, or ECG signals have a different range of values for normal and cardiac patients. Hence, a classifier can be trained with those features for identifying a cardiac patient. It is easy to understand that such a classifier must be supervised in nature and hence it requires a one-time thorough training process to create something called the training model for classification. Remember, a supervised learning-based classifier requires a large amount of data along with the corresponding disease annotations to perform the training operations. In our case, the training data would be recorded as an ECG, PPG, or PCG time series along with the disease labels annotated by doctors.

Preparing the dataset is a very important job for creating a ML classifier. The performance of a classifier largely depends upon the variety of data the classifier learns during training. Hence, one needs to be careful while creating the training data. In practice, the training data should contain a substantial amount of training instances from all the target diseases that the classifier wants to select. For example, if one wants to create a binary classifier for identifying CAD vs non-CAD subject, the CAD population of the training dataset should contain data corresponding to various stages of CAD. This is because the morphology of ECG or PCG varies in different stages of the disease. Moreover, the non-CAD population should not only contain data recorded from healthy subjects, but also consider patient data including data corresponding to cardiac cases other than CAD. The dataset is also ensured to represent a large population demography, age, gender, and BMI in both classes. Opinions from domain experts are taken to determine the inclusion and exclusion criteria of patient selection.

Once the dataset is created, a set of preprocessing jobs are performed on top of the raw sensor data to remove the unusable part of the data generated due to improper placement of the recording device, fixing of the sampling rate of the data, and interpolation of missing data. One may also need to apply signal processing operations such as filtering for removal of the noise components. The signal is now ready for calculation of the relevant features.

The features mentioned in this chapter and in the previous few chapters are typically measured from the time distances between two segments in a signal, e.g., R-R interval distances in an ECG, S1-S2 distances in a PCG. Hence, an accurate segmentation of the signals, such as the extraction of the P wave, the QRS complex, and the T wave form, and the ECG or the S1 and S2 heart sounds regions from a PCG is a prerequisite for feature calculation. It is understandable that accurate segmentation is very important as it directly controls the accuracy of the calculated features in the subsequent stages. Once calculated, the feature values together with the labels are applied to train a classifier. Depending upon the nature of the data and the dimension of the features, a nonlinear SVM or a decision tree-based random forest algorithm are popularly selected as a classifier. Once the classifier is trained, the model is used to predict an unknown subject as either CAD or non-CAD.

Table 1 summarizes a few popular state-of-the-art AI solution designs based upon conventional ML approaches for predicting/classifying heart blockages.

3.2 Deep learning approaches

Machine learning approaches are popular in biomedical applications, but they have one basic limitation. You need to know the discriminating features in advance to train a classifier. The classifier may not give you a good performance if the correct set of discriminating features are not known. Moreover, ML approaches do not perform well on unstructured data and datasets that are noisy. Identifying the optimum set of features is not always easy. Although there are known biomarkers that can be transformed into features for identification of arrhythmias, the same is not true for classification of heart blockages. As mentioned earlier, CAD markers are often subtle at the earliest stage of the disease. For example, there is no single definite clinical marker for identifying a heart blockage. All the features explored in the literature are basically surrogate markers. Moreover, the calculation of features from a large volume of data can be prone to errors. Hence, modern research works are focusing more on a deep learning-based approach, where you just need to feed the classifier with a set of preprocessed and annotated training data. The deep network itself finds the relevant features from the annotated data during training and performs the classification based on them. Modern deep learning approaches are found to outperform the classical ML approaches. However, the training requires much more data, the training model is much bigger in size, and needs a much more complex platform to run. Another major drawback of deep learning approaches lies in the fact that the user only feeds the raw data to the deep learning engine which automatically extracts the relevant features. So, these

Table 1 State-of-the-art heart blockage detection methods using conventional machine learning.

Input signal	Brief methodology	Dataset with description	Accuracy reported
PPG	This method proposes a set of PPG features in time and frequency domain. The features are related to PPG morphology and short term HRV. The features were used to train an SVM classifier for identifying CAD [8].	The methodology was evaluated on an open access dataset and a set of in-house closed dataset recorded by the authors.	The method reported around 90% classification accuracy on both datasets.
	In this method, a subset of 18 features is selected from 40 features based on a wrapper feature selection scheme. These features are extracted from time, frequency, and time-frequency domains. The selected features are fed into an ensemble of 20 feed-forward neural networks for classification [9].	The method was tested on the Open access PhysioNet Challenge 2016 dataset [10], that contains heart sounds corresponding to normal subjects and abnormal heart sounds corresponding to heart blockage and other cardiac conditions.	Around 91% classification accuracy was reported.
PCG	The researchers analyzed a wide list of PCG features in the time and frequency domains along with morphological and statistical features to construct a robust and discriminating feature set for database-agnostic classification of normal and cardiac patients [11].	The methodology was evaluated on the PhysioNet Challenge 2016 dataset.	Reported around 90% classification accuracy on open access datasets

ECG	An automated diagnostic system developed for detection of CAD and myocardial infarction (MI) using (1) discrete wavelet transform (DWT), (2) empirical mode decomposition (EMD), and (3) discrete cosine transform (DCT). ECG signals are subjected to DCT, DWT, and EMD to obtain respective coefficients. These coefficients are reduced using Locality Preserving Projection (LPP) data reduction method. Then, the LPP features are ranked using F-value. Finally, the highly ranked coefficients are fed into the K-nearest neighbor (KNN) classifier [12].	They created their own dataset for evaluation of the algorithm.	This system yielded highest classification results of 98.5% accuracy, 99.7% sensitivity, and 98.5% specificity using only seven features obtained using the DCT technique.

machine-generated features may not have a physical interpretability. Moreover, the performance may decrease on another dataset. Scientists are working on explainable AI to remove these issues. Additionally it has been found that deep learning approaches are capable of finding interesting patterns from noisy data. On the other hand, classification performance of a typical ML classifier such as SVM or a decision tree vastly depends upon the accuracy of the computed features. Feature computation might often not be very accurate on noisy date. For example, in noisy ECG data, the P waves or other low amplitude components are often corrupted by the noise present in the data. The features computed from such data are often not reliable for training an SVM classifier. Deep learning approaches can overcome the issue.

Convolutional neural network (CNN)-based deep learning models are popularly used in medical applications for the analysis of ECG and PCG data for analysis including noise cleaning, disease classification, and anomaly detection. Recently, one-dimensional (1D) CNN has become a powerful tool for the analysis of ECG time series which reportedly outperform conventional ML-based approaches. One-dimensional CNN is known to work well even on noisy ECG data. On the other hand, two-dimensional (2D) CNNs are popularly used with the spectrograms of PCG for analysis.

One-dimensional CNNs are specially designed as a modification of 2D CNN to work on time-series data. Unlike 2D CNN, 1D CNNs have a 1D kernel that moves in a 1D direction across the time series in order to extract temporal features. 1D CNNs are popular in time series classification problems. In most of the applications, raw ECG data are directly fed into a 1D CNN architecture. However, there are applications available in the literature where features such as R-R interval distances are first extracted from the ECG in order to apply to a 1D CNN. As mentioned in Chapter 3, the architecture of a 1D CNN-based deep learning architecture contains the following major blocks:

- Convolutional layer: The input data is applied to a series of layers that perform a 1D convolution operation with a defined kernel size. A specified number of convolution filters are applied in each convolution layer to gradually extract more detailed features that map the input data to a high dimensional feature space. Each convolutional layer is followed by an associated batch normalization layer that normalizes the input data for a faster training experience. Nonlinear activation is applied to the filter output that determines what information to carry forward to the next layer. A rectified linear unit (ReLU) is typically used. ReLU is a piecewise linear function that directly forwards the exact input as output if the input is positive, or returns zero otherwise. ReLU is used as the optimum activation function in most of the CNN architectures as they overcome issues such as vanishing gradients that most of the deep CNN architectures face during training.

- Pooling layer: The convolutional layers contain a number of filters to map the input data to a high dimensional feature space to learn detailed information about the training data. However, this process increases the output dimension of each convolutional layer, which significantly increases the size of number of trainable parameters with redundant information. Pooling layers are applied after each convolution operation to reduce the size of the output feature map.
- Fully connected layer: The output feature map is flattened for final classification. One, or more than one, fully connected layer can be used in CNN architecture.

One-dimensional CNNs have been successfully used in literature for analysis of both single-lead and 12-lead ECGs. Apart from classifying AF and different types of arrhythmias, they have been used for classification of heart blockages and other types of heart diseases. There are also applications where researchers create a preprocessed input from the ECG and apply that to the neural network engine. For example, in the case of classifying AF, where the discriminating marker lies in the irregular pattern in the heart rate variability (HRV), one can extract the R-R intervals from the raw ECG and can apply the same process to the neural network engine in order to learn the desired pattern for classification. One-dimensional CNNs have reported good accuracy in classifying ECG and PPG signals where there are visible morphological changes. For example, there are elevated ST segments of inverted QT interval in the ECG time series which is a clear marker for CAD. However, the situation is different in a PCG. It is difficult to manually classify heart disease from the PCG time series. In most cases, an abnormal signature in a PCG is extra heart sound components at a higher frequency region (typically above 150 Hz). Hence, a spectrogram analysis is typically performed. A spectrogram is a time-frequency representation of time-series data that can be considered as an image. Some literature has reported that 2D CNNs can very accurately classify heart diseases from PCG spectrograms based upon 2D CNN architectures [13,14].

However, ML approaches have their own advantages. In clinical applications, if the features or discriminating markers for identifying a disease are known and they can be easily converted into numerical features, classical ML approaches converge very easily even on a relatively small dataset. This is why the modern deep learning approaches use handcrafted features appended with the deep learning pipeline to create an improved model for better accuracy.

Hence, there is a third hybrid approach, which combines both ML and deep learning. In short, the hand-crafted features are merged with the machine generated features created by the deep network to create a composite feature set which is used for classification. Table 2 lists a few popular state-of-the-art deep learning-based approaches for estimation of heart blockage.

4. Fusion of multiple sensors for classification

In Section 3, we discussed in detail how AI can be successfully used for identifying digital markers that can predict heart blockages. We have also discussed how ML and deep learning-based approaches have been successfully deployed for an accurate estimation of heart blockage. We have also shown some of the advanced approaches published in recent technical forums for prediction of CAD and other types of heart blockages along with the accuracy reported by them. However, it is important to remember that despite the availability of various relevant AI-driven approaches, the classification of CAD is still difficult from a single input source. We have already said in Section 1 that there is no known defined CAD marker present in one signal for a definite prediction of CAD. The features or biomarkers discussed in Section 3 are only indicative surrogate markers that the person is at a high risk of having heart blockage. The presence of these indicative markers is not guaranteed in all stages of CAD. These markers are often inconsistent across the demography. Moreover, many of these markers are often associated with other types of diseases which are not even related to cardiovascular disease. Hence, in order to accurately predict CAD with a higher confidence score, we must rely on multiple independent sensor data. This brings in the concept of sensor fusion. The core idea behind this is, in the case of having multiple nonspecific weak markers to identify a disease, if we can fuse the effect of multiple such markers, we may end up with a more accurate classification performance. A number of recent literature shows that the effect of sensor fusion can actually outperform a single sensor-based approach. For example, PCG alone can be used for accurate detection of CAD and other heart blockages due to presence of abnormal components at higher frequency regions. However, such components are often associated with cardiac

Table 2 State-of-the-art heart blockage detection methods using deep learning techniques.

Input signal	Brief methodology	Dataset description	Accuracy reported
PCG	The approach consists of three major components for classification of heart blockage based on PCG signals: (1) heart sound segmentation, (2) transformation of 1D waveforms into 2D time-frequency heat map representations using Mel-frequency Cepstral coefficients (MFCC), and (3) Classification of MFCC heat maps using deep convolutional neural networks [15].	The approach was evaluated on the PhysioNet Challenge 2016 dataset.	The approach reported around 85% accuracy in identifying abnormal heart sounds.
PCG	This is a hybrid approach combining ML and deep learning. A total of 124 time-frequency features were extracted from the PCG and input to a variant of the AdaBoost classifier. A second classifier using convolutional neural network (CNN) was trained using PCG cardiac cycles decomposed into four frequency bands. The final decision rule to classify normal/abnormal heart sounds was based on an ensemble of classifiers combining the outputs of AdaBoost and CNN [16].	The approach was evaluated on the PhysioNet Challenge 2016 dataset.	Around 92% accuracy was reported on an open access dataset.

Continued

Table 2 State-of-the-art heart blockage detection methods using deep learning techniques—cont'd

Input signal	Brief methodology	Dataset description	Accuracy reported
ECG	In this approach, researchers proposed a classification method of heart diseases based on ECG by adopting a machine learning method, called long short-term memory (LSTM), which is a state-of-the-art technique analyzing time series sequences in deep learning. As suitable data preprocessing. They also utilized symbolic aggregate approximation (SAX) to improve the accuracy [17].	The methodology was reported on the dataset collected by the authors.	Around 90% accuracy was reported.
ECG	Researchers proposed a composite neural network architecture that combined two nonspecific CAD markers: (1) anomalous morphology of ECG waveforms and (2) abnormal HRV. A CNN is defined for extraction of morphological ECG features. Another composite structure is defined based on LSTM and a set of hand crafted statistical features for measuring the extent of HRV. The two independent biomarkers are subsequently combined in a hybrid CNN-LSTM architecture for classification of CAD [18].	Two datasets were used for evaluation. The first dataset was prepared from the open access MIMIC II database [19] and the second database was prepared by the authors at a hospital.	More than 90% classification accuracy was reported on various open access datasets as well as internal datasets recorded by the researchers.

murmur, which is not related to heart blockage. Similarly, HRV features extracted from ECG or PPG can have markers that can lead to other conditions whose signatures are similar to CAD. Hence, if multiple indicative markers from independent sources are found to be present in a person, it can be concluded that the person is more likely to have CAD. Hence, more than one signal containing important biomarkers are combined together to create a fusion-based system. ECG, PCG, and PPG signals can be collectively used in the fusion procedure. All of them are known to carry certain information for identifying CAD. ECG and PPG signals typically contribute to estimating the unique HRV pattern of CAD and identifying the morphological changes in the signal pattern due to cardiac blockage. On the other hand, the abnormalities in opening and closing of heart valves due to a heart blockage can be estimated from a PCG. Hence, the fusion of these signals can produce a better accuracy in prediction of a disease. It can be shown empirically that with the addition of more meaningful sensors, classification accuracy improves significantly.

A typical multisignal decision fusion system with N independent sources is considered to have N independent classifiers that are used at the decision level. N independent classifiers are created based on N different sources. All of them will provide independent predictions. At the final decision layer, a majority voting-based approach can be used for final decision making.

Table 3 lists some existing multimodal approaches for estimation of cardiac blockage.

5. Patient metadata-based knowledge modeling

In Section 4, we discussed how considering more than one weak biomarker can always improve the prediction accuracy. We have also shown how features extracted from ECG, PPG, and PCG can be combined in the classifier for designing a classifier performance. However, there are multiple other factors that directly determine the vulnerability of a person to develop CAD, which is considered a lifestyle disease. A sedentary lifestyle, unhealthy diet, and lack of regular physical activity enhance the likelihood of CAD. Patient demography, and disease history of the patient and their family, play an important role in determining the cardiovascular risk factor of a person. For example, it is found that patients with Type 2 diabetes are at very high risk of developing stroke or CAD [22]. Available research also suggests that a person with a family history of cardiovascular diseases is more vulnerable to forming cardiac blockage. Hence, doctors often emphasize the cardiac

Table 3 State-of-the-art heart blockage detection methods using multiple sensors.

Input signal	Brief methodology	Dataset description	Accuracy
PCG and PPG (both are time-synchronous)	This approach tries to model the human cardiovascular system as a closed loop control system. Being recorded from the source, i.e., the heart, the PCG is considered as the input and the PPG which is recorded from the peripheral body part is considered as the output of the system. The transfer function is measured using a Kalman filter and the filter coefficients are found to be significantly different between a CAD and a non-CAD subject [20].	The method requires both PCG and PPG to be recorded in a time synchronized manner. Hence, getting a dataset is relatively challenging. However, the approach is evaluated on a small dataset reporting very good accuracy.	The method shows that around 90% accuracy is achieved with the combined effect of PCG and PPG.
PCG and PPG (not in time sync)	This approach proposes two unique feature sets extracted from PCG and PPG. Two independent SVM classifiers are created both individually classifying a test subject. If there is a mismatch in the classification result, the output of the classifier with a better confidence score is taken as final [21].	Tested on a small in-house dataset, results show that the multimodal approach outperforms the individual accuracy produced by the PCG and the PPG based classifiers.	100% accuracy was achieved on the small dataset.

disease history of a patient for critical decision making. Additionally, a number of other relevant parameters such as blood pressure, cholesterol level, body weight, BMI, fasting glucose level, and whether a person is a smoker or not, are also taken into consideration when consulting with a patient. One or more "yes" answers to these questions may categorize the patient as a high risk even if they do not show any symptoms at the time of visiting the clinic. All this information, collectively known as clinical metadata, can also be used in the AI-based decision support system to improve the performance of a CAD classifier. In Ref. [23], the authors have shown that CAD can be accurately predicted using ML using various clinical attributes including patient medical data. The approach was successfully evaluated on a dataset which is openly available and is provided by the University of California Irvine (UCI) Machine Learning Repository. This data contains 303 instances with a total of 14 clinical attributes.

Pure ML-based heart blockage classifiers also use this information as an additional knowledgebase to improve their classification performance. The clinical metadata are used to create a hierarchical tree-based system to identify the high-risk patients. The clinical metadata information are a set of values combining numerical numbers (e.g., blood glucose level, blood pressure, etc.) or categorical parameters (e.g., whether the subject is a smoker or has a past cardiac disease history). Part of this metadata can be obtained via questionnaire and the rest via periodic health checkups. A combination of these parameters can be used to determine whether the subject is at high risk. A multisignal fusion-based classifier can make a more effective classifier for identifying cardiac patients.

Fig. 3 summarizes the research work proposed in Ref. [24] that devised a two-stage approach for prediction of CAD. As shown in the figure, the first stage comprises a rule engine based on patient lifestyle and disease history, which coarsely estimates the risk factor of the subject of interest. The high-risk patients undergo a detailed analysis in the second stage based on a multisignal fusion approach based on numerical handcrafted features extracted from the ECG, PCG, and PPG. The combined feature set obtained in the process are applied to a SVM classifier for the final prediction. The authors in Ref. [17] undertook a detailed analysis of the impact of the two-stage approach on 150 hospital patients. The results revealed that the two-stage approach not only outperforms the benchmark automated CAD detection tools, but also is more efficient in mass screening as only a small fraction of the test population is required to undergo the second stage. The method was evaluated on 150 real-world patients in an urban hospital

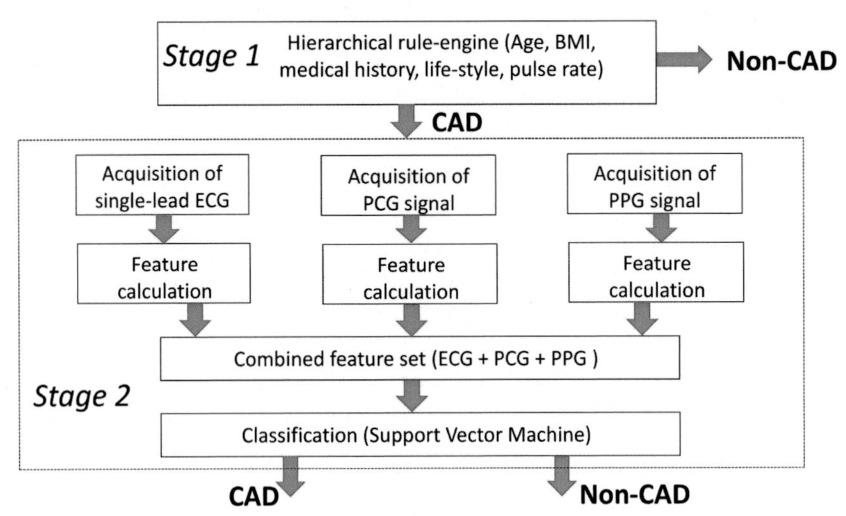

Fig. 3 Two-stage approach for detection of CAD.

setup. The data was recorded using commercially-available, low-cost, unobtrusive sensors. The results show that the proposed two-stage approach is not only better than the existing approaches that only rely on biomedical signals, but also is very efficient in screening the high-risk patients from a population. In general, the method yields around 95% accuracy in classifying CAD patients.

6. Conclusion

Cardiovascular diseases are considered to be silent killers. A blockage in the coronary arteries may restrict the blood flow which eventually leads to a stroke or cardiac arrest. Millions of people die every year due to complications caused by heart blockages. Doctors strongly believe that the fatal consequences of a cardiac disease can often be prevented if it is detected at an early stage. However, cardiovascular diseases are often asymptomatic in their early stages. This requires continuous monitoring, possibly 24 × 7 monitoring for high-risk patients. AI can play a crucial role in unobtrusive continuous patient monitoring and timely alert generation. In this chapter, we have shown that critical markers for identifying a heart blockage can be found in noninvasive biomedical signals such as ECG, PCG, or PPG. We have shown different markers available in noninvasive biosignals have been found to carry important information for the early prediction of CAD. We have further discussed how modern AI, ML, and deep learning can be used for the prediction of heart blockage based on these signals. We have briefly covered a few leading research articles that have reported promising accuracy in identifying heart blockages. We have also shown how a strong classification performance can be achieved via fusion of multiple biosignals. However, it is to be remembered that the performance of the majority of the deep learning-based solutions strongly depends upon the diversity of the dataset upon which it has been trained. The test data is expected to be capturing similar properties to that of the training data and a ML algorithm is not guaranteed to work on totally unseen data. Hence, one must critically select the training dataset when creating a ML or a deep learning-based model. Further, despite yielding an accuracy which is close to 100%, the AI-based solutions are only to be used for screening or prediagnosis purposes to assist clinicians. They are not designed to replace the gold standard clinical diagnosis tools popularly used in hospitals for the actual diagnosis of heart diseases.

Finally, we remind you that the accurate classification of heart blockage including CAD using AI is still a very much open area of research. Available applications have mostly been reported on relatively small datasets and the data were mostly recorded from a particular demography. However, the prevalence of heart blockage depends upon demography and lifestyle. Hence, the AI models trained on one particular group are likely to fail on another population.

References

[1] S. Virani, et al., Heart disease and stroke statistics—2021 update: a report from the American Heart Association, Circulation 143 (8) (2021).

[2] S. Nakamura, T. Uzu, T. Inenaga, G. Kimura, Prediction of coronary artery disease and cardiac events using electrocardiographic changes during hemodialysis, Am. J. Kidney Dis. 36 (3) (2000) 592–599.

[3] S. Dua, S. Xian Du, V. Sree, V.I. Thajudin Ahamed, Novel classification of coronary artery disease using heart rate variability analysis, J. Mech. Med. Biol. 12 (04) (2012) 1240017.

[4] G. Altan, Diagnosis of coronary artery disease using deep belief networks, Eur. J. Eng. Nat. Sci. 2 (1) (2017) 29–36.

[5] G. Marcus, J. Vessey, M.V. Jordan, M. Huddleston, B. McKeown, I.L. Gerber, E. Foster, K. Chatterjee, C.E. McCulloch, A.D. Michaels, Relationship between accurate auscultation of a clinically useful third heart sound and level of experience, Arch. Intern. Med. 166 (6) (2006) 617–622.

[6] M. Akay, J.L. Semmlow, W. Welkowitz, M.D. Bauer, J.B. Kostis, Detection of coronary occlusions using autoregressive modeling of diastolic heart sounds, IEEE Trans. Biomed. Eng. 37 (4) (1990) 366–373.

[7] M. Akay, M. Bauer, J.L. Semmlow, W. Welkowitz, J. Kostis, Autoregressive modeling of diastolic heart sounds, in: Proceedings of the Annual International Conference of the IEEE Engineering in Medicine and Biology Society, IEEE, 1988, pp. 172–174.

[8] R. Banerjee, K.M. Ramu Vempada, A.D. Mandana, Choudhury, and Arpan Pal., Identifying coronary artery disease from photoplethysmogram, in: Proceedings of the 2016 ACM International Joint Conference on Pervasive and Ubiquitous Computing: Adjunct, 2016, pp. 1084–1088.
[9] M. Zabihi, A.B. Rad, S. Kiranyaz, M. Gabbouj, A.K. Katsaggelos, Heart sound anomaly and quality detection using ensemble of neural networks without segmentation, in: 2016 Computing in Cardiology Conference (CinC), IEEE, 2016, pp. 613–616.
[10] G.D. Clifford, C. Liu, B. Moody, D. Springer, I. Silva, Q. Li, R.G. Mark (Eds.), Classification of normal/abnormal heart sound recordings: the PhysioNet/Computing in Cardiology Challenge 2016, 2016 Computing in Cardiology Conference (CinC), IEEE, 2016, pp. 609–612.
[11] R. Banerjee, A.D. Choudhury, P. Deshpande, S. Bhattacharya, A. Pal, K.M. Mandana, A robust dataset-agnostic heart disease classifier from phonocardiogram, in: 2017 39th Annual International Conference of the IEEE Engineering in Medicine and Biology Society (EMBC), IEEE, 2017, pp. 4582–4585.
[12] U.R. Acharya, H. Fujita, O. Muhammad Adam, S. Lih, V.K. Sudarshan, T.J. Hong, J. E.W. Koh, et al., Automated characterization and classification of coronary artery disease and myocardial infarction by decomposition of ECG signals: a comparative study, Inform. Sci. 377 (2017) 17–29.
[13] Z. Ren, N. Cummins, V. Pandit, J. Han, K. Qian, B. Schuller, Learning image-based representations for heart sound classification, in: Proceedings of the 2018 International Conference on Digital Health, 2018, pp. 143–147.
[14] Y. Chen, S. Wei, Y. Zhang, Classification of heart sounds based on the combination of the modified frequency wavelet transform and convolutional neural network, Med. Biol. Eng. Comput. 58 (9) (2020) 2039–2047.
[15] J. Rubin, R. Abreu, A. Ganguli, S. Nelaturi, I. Matei, K. Sricharan, Classifying heart sound recordings using deep convolutional neural networks and mel-frequency cepstral coefficients, in: 2016 Computing in Cardiology Conference (CinC), IEEE, 2016, pp. 813–816.
[16] C. Potes, S. Parvaneh, A. Rahman, B. Conroy, Ensemble of feature-based and deep learning-based classifiers for detection of abnormal heart sounds, in: 2016 Computing in Cardiology Conference (CinC), IEEE, 2016, pp. 621–624.
[17] M. Liu, Y. Kim, Classification of heart diseases based on ECG signals using long short-term memory, in: 2018 40th Annual International Conference of the IEEE Engineering in Medicine and Biology Society (EMBC), IEEE, 2018, pp. 2707–2710.
[18] R. Banerjee, A. Ghose, K.M. Mandana, A hybrid CNN-LSTM architecture for detection of coronary artery disease from ECG, in: 2020 International Joint Conference on Neural Networks (IJCNN), IEEE, 2020, pp. 1–8.
[19] M. Saeed, M. Villarroel, A.T. Reisner, G. Clifford, L.-W. Lehman, G. Moody, T. Heldt, T.H. Kyaw, B. Moody, R.G. Mark, Multiparameter Intelligent Monitoring in Intensive Care II (MIMIC-II): a public-access intensive care unit database, Crit. Care Med. 39 (5) (2011) 952.
[20] R. Banerjee, A.D. Choudhury, S. Datta, A. Pal, K.M. Mandana, Non invasive detection of coronary artery disease using PCG and PPG, in: eHealth 360°, Springer, Cham, 2017, pp. 241–252.
[21] A.D. Choudhury, R. Banerjee, A. Pal, K.M. Mandana, A fusion approach for non-invasive detection of coronary artery disease, in: Proceedings of the 11th EAI International Conference on Pervasive Computing Technologies for Healthcare, 2017, pp. 217–220.
[22] J.-Y. Han, D.-H. Choi, S.-W. Choi, B.-B. Kim, Y.-J. Ki, J.-W. Chung, Y.-Y. Koh, K.-S. Chang, S.-P. Hong, Stroke or coronary artery disease prediction from mean platelet volume in patients with type 2 diabetes mellitus, Platelets 24 (5) (2013) 401–406.

[23] A.G. Yallappagoudar, C.M. Arun, S. Gangadhar, V. Agarwal, M. Anitha, Prediction of coronary artery disease using machine learning, Int. J. Recent Adv. Multidiscip. Topics 1 (3) (2020) 4–7.

[24] R. Banerjee, A. Ghose, A. Sinha, A. Pal, K.M. Mandana, A multi-modal approach for non-invasive detection of coronary artery disease, in: Adjunct Proceedings of the 2019 ACM International Joint Conference on Pervasive and Ubiquitous Computing and Proceedings of the 2019 ACM International Symposium on Wearable Computers, 2019, pp. 543–550.

Further reading

R. Walsh, Hurst's the Heart Manual of Cardiology, McGraw-Hill Education, 2012.

G. Coppola, P. Carità, E. Corrado, A. Borrelli, A. Rotolo, M. Guglielmo, C. Nugara, L. Ajello, M. Santomauro, S. Novo, ST segment elevations: always a marker of acute myocardial infarction? Indian Heart J. 65 (4) (2013) 412–423.

R. Jain, R. Singh, S. Yamini, M.K. Das, Fragmented ECG as a risk marker in cardiovascular diseases, Curr. Cardiol. Rev. 10 (3) (2014) 277–286.

D. Velusamy, K. Ramasamy, Ensemble of heterogeneous classifiers for diagnosis and prediction of coronary artery disease with reduced feature subset, Comput. Methods Programs Biomed. 198 (2021) 105770.

R. Alizadehsani, J. Habibi, M.J. Hosseini, R. Boghrati, A. Ghandeharioun, B. Bahadorian, Z.A. Sani, Diagnosis of coronary artery disease using data mining techniques based on symptoms and ECG features, Eur. J. Sci. Res. 82 (4) (2012) 542–553.

M.C. Çolak, C. Çolak, H. Koçatürk, Ş. Sağıroğlu, İ. Barutçu, Predicting Coronary Artery Disease Using Different Artificial Neural Network Models, 2008.

G. Jackson, N. Boon, I. Eardley, M. Kirby, J. Dean, G. Hackett, P. Montorsi, et al., Erectile dysfunction and coronary artery disease prediction: evidence-based guidance and consensus, Int. J. Clin. Pract. 64 (7) (2010) 848–857.

R. Alizadehsani, M.J. Hosseini, A. Khosravi, F. Khozeimeh, M. Roshanzamir, N. Sarrafzadegan, S. Nahavandi, Non-invasive detection of coronary artery disease in high-risk patients based on the stenosis prediction of separate coronary arteries, Comput. Methods Programs Biomed. 162 (2018) 119–127.

W.C. Little, M. Constantinescu, R.J. Applegate, M.A. Kutcher, M.T. Burrows, F.R. Kahl, W.P. Santamore, Can coronary angiography predict the site of a subsequent myocardial infarction in patients with mild-to-moderate coronary artery disease? Circulation 78 (5) (1988) 1157–1166.

P.W.F. Wilson, Established risk factors and coronary artery disease: the Framingham Study, Am. J. Hypertens. 7 (7_Pt_2) (1994) 7S–12S.

O. Honda, S. Sugiyama, K. Kugiyama, H. Fukushima, S. Nakamura, S. Koide, S. Kojima, et al., Echolucent carotid plaques predict future coronary events in patients with coronary artery disease, J. Am. Coll. Cardiol. 43 (7) (2004) 1177–1184.

M. Inouye, G. Abraham, C.P. Nelson, A.M. Wood, M.J. Sweeting, F. Dudbridge, F.Y. Lai, et al., Genomic risk prediction of coronary artery disease in 480,000 adults: implications for primary prevention, J. Am. Coll. Cardiol. 72 (16) (2018) 1883–1893.

L. Verma, S. Srivastava, P.C. Negi, A hybrid data mining model to predict coronary artery disease cases using non-invasive clinical data, J. Med. Syst. 40 (7) (2016) 1–7.

A.C. Dortimer, R.L. DeJoseph, R.A. Shiroff, A.J. Liedtke, R. Zelis, Distribution of coronary artery disease. Prediction by echocardiography, Circulation 54 (5) (1976) 724–729.

CHAPTER 6

Hypertension and cardiac fatigue

1. Introduction

When blood flows, it exerts a certain pressure against the wall of the arteries. This is called blood pressure. As the intensity of blood volume through the arteries is not fixed in a cardiac cycle, the blood pressure also varies accordingly. It increases during systole to reach a maximum value and decreases during diastole to reach a minimum value. The maximum and the minimum value of the blood pressure are known as systolic and diastolic blood pressure respectively, and the difference is termed pulse pressure. A systolic blood pressure (SBP) of 120 mmHg and diastolic blood pressure (DBP) of 80 mmHg is usually termed as normal. Blood pressure may elevate from the normal range during exercise, physical activities, or due to mental stress. A persistent condition of abnormally high blood pressure is termed hypertension. According to the World Health Organization (WHO), hypertension is a serious medical condition that can be the underlying cause for an increase in the risk of heart, brain, and kidney diseases. It is well known that a major proportion of premature deaths worldwide are initiated by uncontrolled hypertension, with upwards of 1 in 4 men and 1 in 5 women, amounting to a total of over a billion people suffering from the condition. It should also be noted that the burden of hypertension is disproportionately high in low- and middle-income countries, which accounts for two-thirds of the global cases. This is largely due to increased risk factors in those populations in recent decades [1,2]. The primary reasons for developing hypertension are uncontrolled body weight due to unhealthy lifestyle such as lack of physical exercise, consumption of too much junk food, excessive smoking or drinking, and lack of sound sleep. Additionally, genetics also plays an important role here as it is well established that those who with a family history of hypertension are more likely to develop the condition. This genetic effect is elaborated more in Chapter 8.

Because blood flow through arteries is a continuous process, as the heart continuously pumps blood, blood pressure can have different values depending upon the blood volume flowing through the artery. However,

New Frontiers of Cardiovascular Screening using Unobtrusive Sensors, AI, and IoT
https://doi.org/10.1016/B978-0-12-824499-9.00006-4

there are two particular values of blood pressure that interest clinicians: the systolic value and the diastolic value. The systolic value is the maximum pressure value exerted by the blood against the walls of the artery when the heart beasts. The diastolic value is the pressure exerted by the blood against the arterial walls when the heart relaxes. The difference between systolic and diastolic pressure is termed as the pulse pressure. However, in common terms, blood pressure is represented as systolic pressure/diastolic pressure in mmHg unit. The typical blood pressure of a normal person is considered to be 120/80 mmHg. Five different ranges of blood pressure are recognized by the American Heart Association. A systolic value less than 120 mmHg and diastolic blood pressure below 80 mmHg are considered normal. Most young healthy adults fall under this category. However, blood pressure may slightly elevate with age and may be lesser for women. Systolic blood pressure in the range of 120–129 mmHg along with diastolic blood pressure less than 80 mmHg causes a condition termed elevated blood pressure. Persons who have higher than normal blood pressure are likely to develop hypertension if left untreated. A further enhancement in blood pressure with systolic value in the range of 130–139 mmHg and diastolic value in the range of 80–89 mmHg causes a condition called Stage 1 hypertension. In this stage the doctors suggest lifestyle changes or prescribe medicines because the patient has a significantly higher risk of developing cardiovascular diseases. In Stage 2 hypertension, the blood pressure consistently ranges above 140/90 mmHg. At this stage the doctors prescribe medicines and restrictions are imposed on patient's lifestyle and diet. There is another condition called low blood pressure or hypotension, which occurs when the blood pressure of a person consistently sits below 90/60 mmHg. The brain does not receive enough blood which causes fainting or dizziness. Persons with anemia are likely to form this condition which may require medication. Just like any other cardiovascular diseases, once formed, hypertension cannot be cured. However, medication may help in controlling blood pressure. It is very important to determine the onset of hypertension. A lifestyle change with regular workouts and appropriate diet can help in controlling the blood pressure with minimum medication. Although hypertension affects most people in their forties, in recent times people are becoming affected at a much younger age. Hence, doctors advise regular monitoring of blood pressure. A cuff-based mercury sphygmomanometer is the clinically trusted equipment for measuring blood pressure. However, the process is not unobtrusive and requires the assistance of medical technician to listen to the sound for measuring blood pressure. Hence, it cannot be used in household for day-to-day monitoring. Digital blood pressure monitoring tools have been popular for this purpose. Despite being less accurate compared to an analog mercury device, they are user-friendly and can be used at home. However, it is still very similar to the cuff-based measurement. Nowadays, noninvasive cuffless blood pressure monitoring is being investigated. The objective behind this is to design a fully automated unobtrusive blood pressure monitoring tool that can be used for continuous monitoring that is very useful for hypertension patients. These tools can not only monitor the blood pressure 24×7 even during various activities, but also can generate alerts when the blood pressure elevates.

Photoplethysmogram (PPG) signals measure the blood volume in blood vessels, and it is widely believed to be an important source for estimating blood pressure. Research shows that biomedical signals such as a PPG, which is a measurement of blood volume in capillaries, can be used as an indirect marker for predicting blood pressure. It is to be noted that, unlike the medical-grade mercury sphygmomanometer which measures the blood pressure, a PPG signal can only be used to predict the blood pressure indirectly with some accuracy. However, its utility does not lie in very accurately measuring blood pressure. It is mostly used in the form of a continuous monitoring tool in the form factor of a wearable device that can track hypertension. In this chapter, we will discuss in detail how AI can be used to indirectly estimate blood pressure using easy-to-record signals for the measurement of blood pressure. Initially, we will discuss how blood flow is related to blood pressure. We will use a simple yet popular mathematical model which is analogous to the human cardiovascular system. We will relate blood flow to blood pressure using a set of equations. Subsequently, we will show how PPG signals can be used to measure blood pressure. Finally, we will introduce a powerful concept called the pulse transit time (PTT) which is measured from the peak to peak interval distances of PPG and ECG signals recorded in a time-synchronous manner. PTT is considered a parameter for very accurate measurement of blood pressure.

In this chapter, we will also discuss another important parameter cardiac fatigue which is a direct cause of sudden cardiac arrest. Just like working too long causes tiredness, stressing the heart too much can also cause cardiac fatigue. Every heart has a limited capacity and going beyond that may tire the heart which may cause a sudden cardiac death.

2. Screening of hypertension from PPG and ECG

The mercury sphygmomanometer is considered the globally acknowledged tool for accurate measurement of the blood pressure. The tool comes with a cuff and a rubber bladder and requires the help of a medical technician to measure the blood pressure of a person. The measurement is done from the brachial artery of the upper arm. This artery is the continuation of the axillary artery beyond the lower margin of the teres major muscle. The cuff is first wrapped around the upper arm of the person and the rubber bladder is inflated to exert high pressure to squeeze the artery to stop the blood flow. Subsequently, the pressure is released slowly, and the technician waits for a sound termed the Korotkoff sound, which indicates the blood has just started flowing through the artery. The corresponding pressure value is noted as a systolic value. The Korotkoff sound is heard using a stethoscope placed on the arm close to the artery. There is mercury bulb attached with a scale to measure the pressure level. Once the systolic pressure is noted, the pressure is further reduced to calculate the diastolic pressure. Despite their accuracy, mercury sphygmomanometers are not for home measurement as it requires human assistance with expertise to identify the Korotkoff sound. Digital blood pressure monitors have become very popular. It uses a combination of inflatable air-bladder cuff, a battery-powered air pump, and a pressure sensor for sensing arterial wall vibrations to effectively measure the arterial blood pressure. Although less accurate compared to a mercury sphygmomanometer, they are easy to use and hence are popular because it does not require a medical technician to measure blood pressure. However, it still cannot be used for continuous beat-by-beat monitoring of blood pressure. Artificial intelligence (AI) = based approaches have been successful for continuous blood pressure monitoring which can indicate the onset of hypertension. It is to remember that that AI approaches are for estimation of blood pressure, not for its measurement.

In this section we will discuss how various noninvasive signals can be used for the screening of hypertension. We will start with some popular electrical models that mimic the analogy of the human cardiovascular system to relate blood flow and blood pressure. Later we will show how PPG and ECG can be used to predict blood pressure. In order to do so, we will introduce the PTT which can be measured from the PPG and the ECG, and which is directly related to blood pressure.

2.1 Electrical modeling of cardiovascular system

An electrical model popularly refers to a lumped model that is analogous to the human cardiovascular system. Lumped models are models that mathematically formulate the load to the heart, where the entire systemic arterial

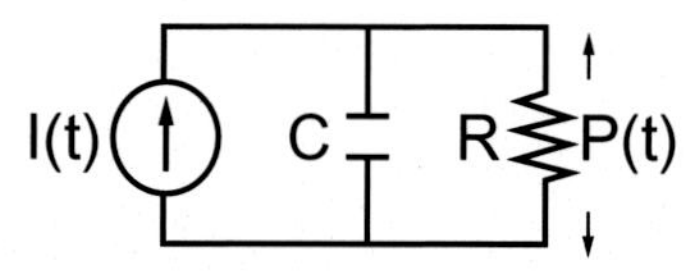

Fig. 1 The two-element Windkessel model.

tree is represented as lumped electrical components such as resistors, capacitors, and inductors.

In 1733, Stephen Hales first introduced a simple yet conceptual model of the arterial tree. According to his theory, the interaction between the heart and the arteries is similar to the working principle of a fire hose, in which the pulsatile action of the pump is damped by an air chamber. The word "Windkessel" loosely translates the word "air chamber" in German. Hence, the model is termed the Windkessel model. In the cardiovascular system, large arteries play the role of the air chamber [3,4].

In 1899, Otto Frank refined and mathematically formulated the idea by introducing the two-element Windkessel model, the simplest form of the lumped electrical model. A circuit diagram of the two-element Windkessel model is shown in Fig. 1.

As the name suggests, the two-element Windkessel model comprises two lumped components, a resistance (R) and a capacitance (C). The resistance is analogous to the peripheral resistance offered by the arteries against the blood flow and the capacitance is analogous to the vessel compliance. The circuit is powered by the AC current source $I(t)$, which is analogous to the blood pumped by the heart. The output voltage $P(t)$ is analogous to the blood pressure caused due to blood flow. Both $I(t)$ and $P(t)$ are time varying.

In electrical circuits, Kirchhoff Laws are popularly used which is a set of two differential equations (the current law and the voltage law) that deal with the current and potential difference in the lumped element model of electrical circuits.

Applying the Current law on the circuit, we get:

$$\frac{P(t)}{R} + C\frac{dP(t)}{dt} = I(t)$$

The cardiac cycle (T_c) is a sum of two components, the systolic time (T_s) and the diastole time (T_d). The heart only pumps blood during systole. Typically, the diastole time is 1.5 times more than the systole time. In its simplest form, the arterial blood flow ($I(t)$ in the Windkessel model) can be simply defined as a rectified sinusoidal curve in a cardiac cycle as follows:

$$I(t) = I \sin \frac{\pi t}{T_s} \quad \text{for } 0 < t \leq T_s$$

$$I(t) = 0 \quad \text{for } T_s < t \leq T_c$$

Here, the maximum blood flow is termed I, which typically occurs at the middle of systole. The value of I is constant, which can be measured from cardiac output. The cardiac output is the measure of total blood volume pumped by the heart in a minute. If the cardiac output is Co, the blood flow in one cardiac cycle is:

$$\frac{Co \cdot 60}{T_c}$$

The whole term is the equal to the integration of the term $I(t)$. Hence:

$$\frac{Co \cdot 60}{T_c} = \int_0^{T_c} I \sin \frac{\pi t}{T_s}$$

Now, blood pressure $P(t)$ can be solved by putting the expression of $I(t)$ in the first differential equation and solving the same. Solving that, we calculate the systolic (P_s) and the diastolic (P_d) blood pressure values in terms of I, R, C, T_s, and T_d as:

$$P_s = P_d e^{\frac{-T_s}{RC}} + \frac{I \cdot T_s \cdot C \cdot \pi \cdot R^2}{T_s^2 + C^2 \cdot \pi^2 \cdot R^2}\left(1 + e^{\frac{-T_s}{RC}}\right)$$

$$P_d = P_d e^{\frac{-T_d}{RC}}$$

The above expressions are the simplest form of blood pressure values solved using the two-element Windkessel model. The human cardiovascular system is very complex in nature. Hence, modeling it using a single resistance and capacitance is not accurate. Consequently, modified versions of Windkessel models have been subsequently developed. Improved versions of the simple two-element Windkessel model are three- and four-element model as shown in Fig. 2A and B.

In a three-element Windkessel model, apart from the peripheral resistance (R_p) and vessel compliance (C), a new resistance (R_a) is introduced which represents the resistance faced by the blood while entering the aortic or pulmonary valve.

The four-element Windkessel model is the most advanced version of the model which extends the three-element model by adding an inductance (L_c) to represent the inertia of the blood.

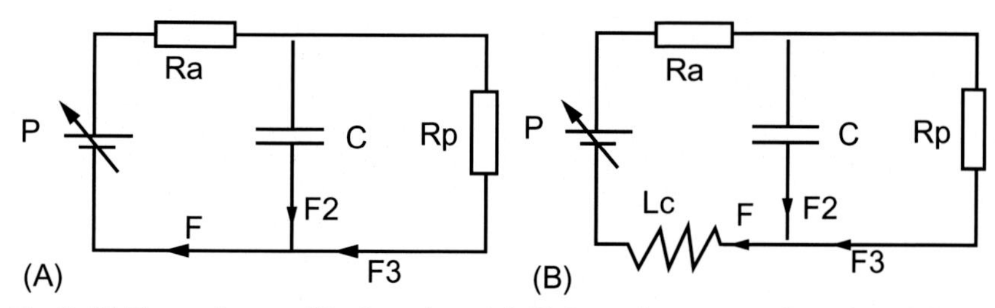

Fig. 2 (A) Three-element Windkessel model. (B) Four-element Windkessel model.

More complex electrical models have been developed [5,6] to accurately simulate the blood pressure that considers most of the functions of the cardiovascular system. However, the Windkessel model still remains a pioneering method.

In this chapter we have discussed lumped electrical models to relate between blood flow and blood pressure. We have also derived a set of mathematical equations to estimate blood pressure if the blood flow is known. However, it is important to remember that getting the actual volume of blood volume is difficult, and it is impossible to get in a pure noninvasive manner. The AI-based blood pressure estimators thus rely on noninvasive, easy-to-measure signals that can coarsely estimate the blood flow in order to measure BP in an empirical manner. PPG signals measure the peripheral blood flow in capillaries and are widely used in existing AI-enabled systems for the estimation of BP. PPG is easy to measure and can be measured from the fingertip using a pulse oximeter or can even be recorded using a smartphone. Thus, PPG-based systems do not require a cuff to wrap around the arm can be used to predict beat-to-beat blood pressure for continuous monitoring. This helps in locating the fluctuation in blood pressure which is considered an early marker for hypertension.

2.2 Other lumped models for simulation of arterial blood pressure

The finite volume simulation of blood flow is a special instance of biomedical simulations. There are modern research works for pressure flow simulation using numerical models that can be used for generating arterial pressure flow. Blood can be considered a Newtonian fluid in mechanical models and hence can be described by the popular Navier-Stokes equations or by particle methods [7]. Modern lumped electrical models use a complex structure to simulate individual valves and arteries in the heart to create a detailed view of the cardiovascular system. In Ref. [8] the authors proposed a numerical model for this purpose. On the other hand, there are fluid flow-based mechanical models for simulation of arterial blood pressure [9,10]. Hemodynamic models have also been popular in recent years. Hemodynamics is the study of blood flow, where different parameters are mathematically defined to quantify blood flow and the relationship with systemic circulatory changes in form of a set of differential equations. In Ref. [11], the authors proposed a 1D hemodynamic model to simulate the coronary artery tree. In Ref. [12], a personalized model based on combination of 0D and 1D hemodynamic models was proposed to simulate the human cardiovascular system. Hemodynamic modeling of human cardiovascular system is still an open area of research.

2.3 Regression modeling from PPG

Blood flow causes blood pressure. In the previous section we have seen how blood pressure is related to blood flow in terms of a few lumped parameters as well as the duration of cardiac cycle. An accurate estimation blood volume is a nontrivial task. There is no known invasive method for measurement. However, from the PPG signal we can estimate a few parameters to indirectly understand important features of the net blood volume. The blood pumped by the heart is branched through various arteries. The blood flow in the peripheral capillaries can be measured in terms of PPG. However, a number of parameters such as cardiac cycle length, systole time, and diastole time can be measured from the PPG. A number of research works have shown that numerical features extracted from the PPG can be used to predict the blood pressure using a machine learning (ML)-based technique. For that purpose, a regression network is trained. Unlike a classifier, a regression network can predict a target value rather than predicting a discrete target class. We can think of linear regression as the simplest form of a regression network. For example, let us consider x as an independent variable that is related to the dependent variable y as follows:

$$y = m \cdot x + c$$

Recall, this is the expression of a straight line where m is called the slope and c is called the intercept. Suppose we do not know m and c but we have a set of x and the corresponding y. In that case, m and c can be predicted using a linear regression approach. In this method, an initial value of m and c is assumed and the value of predicted y (let's call it y') is measured. Now the mean square error between the actual y and the y' is minimized. When that sum is reached, the network is considered to be trained, and the value of m can be empirically learned. Now, a new set of x and y can be predicted. The independent variable x can be multivariate.

In our case we can assume x as our PPG feature and y is the predicted blood pressure value. In a practical scenario, x and y can have a complex nonlinear relationship. Hence a linear regression can be an underfit. There are nonlinear regression techniques as well, using, for example, an artificial neural network (ANN), or support vector machine regressor (SVR).

Lamonaca et al. [13] proposed a novel feature set from PPG for the estimation of BP. The approach comprises features that measure the width of the PPG pulse at different pulse heights. A total of 20 features were derived related to pulse width at a given pulse height, the rising and falling time of PPG and the slopes. An artificial neural network-based regression model is trained using the features for estimation of BP. The approach gives around 10 mmHg error in prediction on a public dataset in prediction of blood pressure. A number of studies suggest that demographical information can have an important role in BP. The normal BP range can be elevated for a person with a higher BMI. In another approach, PPG features are added with demographic information such as height, weight, and age to derive an improved feature set for the prediction of BP [14]. The features are used to train a SVR for prediction of BP. Their result suggests that inclusion of demography information can improve the prediction performance. The method was designed to effectively run on smartphone where the PPG is captured using the inbuilt flash and camera of the device. This approach suggests that blood pressure can be successfully estimated even on a smartphone without purchasing a dedicated hardware. However, PPG signals captured in a smartphone are of inferior quality and require a detailed noise cleaning for an accurate feature extraction. Deep learning approaches have also been successfully used recently. In Ref. [15], the authors proposed a CNN-based blood pressure estimator from PPG. The method is capable of detecting the small fluctuation of BP happening within a small period of time. In Ref. [16], the authors proposed a long short-term memory (LSTM) network for the prediction of BP. All these approaches are capable of estimating beat-by-beat blood pressure and hence can be used for continuous monitoring.

The abovementioned approaches are purely data driven. Their objective is to extract a set of features from the PPG and other relevant sources to train a ML model that can predict BP. The performance of the ML model solely depends upon the diversity of the training database. These approaches are often not reliable to predict very high blood pressure which may lead to hypertension. The electrical model approaches can be a solution to address this issue. The core idea is to relate the blood flow from PPG so that BP can be subsequently measured using mathematical equations.

A newer method of deriving blood pressure from PPG signals was proposed in the research work in Ref. [17] that utilizes an effective combination of an electrical model and a data-driven model. In the said approach, first a set of features were extracted from PPG. However instead of generating a regression equation for relating them to the blood pressure value, the authors first related the resistance and the capacitance parameter of the two-element Windkessel model using linear regression-based curve-fitting, which are considered to be the latent parameters. Once the latent parameters are estimated, the set of equations discussed in the previous sections are used for the estimation of BP. The authors have shown that the said approach outperforms the previously developed pure data-driven approaches on an open access dataset. A refined version of the algorithm was later proposed by the authors [18], that overcomes a few limitations of the original approach. The modified approach not only considers the demographic information of the users, but also mathematically derives a number of parameters that were previously fixed at a normal value. The modified approach was tested on two datasets recorded by the authors using a low-cost pulse oximeter. The algorithm could predict blood pressure accurately with less than 8 mmHg error. In Ref. [19], the authors modified their approach to work on potentially noisy PPG data recorded by a smartphone. The authors showed that a PPG signal can be modeled using a sum of two Gaussian curves and the parameters of the Gaussian curves can be used as features to train the latent parameters for estimating blood pressure. The second approach can successfully be used to predict BP even on a smartphone that records noisy PPG signals.

Over the last decade a lot of work has been undertaken for indirect estimation of BP. PPG has been considered the primary source for noninvasive prediction of BP. However, pure PPG-based solutions are not often very accurate. Moreover, PPG signals are vulnerable to background noise. The extraction of relevant features from a noisy PPG can be erroneous which can lead to inaccurate prediction problem. In the next section, we will talk about a new methodology that can result in a very accurate BP estimation using a limited number of features.

3. Pulse transit time analysis from PPG and ECG

So far, we have discussed the relationship between blood flow and blood pressure. We have discussed a simple yet powerful electrical model for mathematically relating the two. Although the Windkessel model is an oversimplified way of representing the human cardiovascular system, it has inspired a number of modern powerful models in use. We have also discussed noninvasive and unobtrusive estimation of blood pressure. We have used the PPG signal for that purpose. Available research works show that the PPG signal can be reliably used for estimation of blood pressure. However, we have not talked much about the accuracy of those approaches. In the Windkessel model, we assume the heart is the source of pumping blood which causes arterial blood pressure. However, in practice, when we estimate blood pressure, we do not have access to the actual volume of the blood pumped by the heart. What we have is the PPG signal, which is the blood volume measured at the peripheral body parts such as the fingertip or the toe. The blood pumped by the heart is branched through various arteries before eventually reaching the peripheral body parts. Hence, it is very difficult to estimate the actual characteristics of the blood flow from PPG. Thus, the blood pressure monitoring techniques which take PPG as their sole input may not be sufficient to capture the blood pressure very accurately as the PPG signal contains very limited information. In practice, various state-of-the-art approaches have reported that the PPG-based techniques are enough to accurately measure the blood pressure when it falls in the normal range. However, they often fail to correctly estimate high blood pressure. Moreover, they are not fruitful in estimating the rapid fluctuation in blood pressure within a short time interval. Thus, they have limitations in deployment for monitoring hypertensive and prehypertensive patients.

There is another parameter, called the pulse transit time (PTT) [20], which is popularly used for estimating blood pressure. PTT is defined as the time taken for the arterial pulse pressure wave to travel from the aortic valve to a peripheral site. In general, it is a measure of the flow velocity which can be measured from the time difference of the aortic wave between reaching two different parts of the human body. PTT is used to measure a parameter, called the pulse wave velocity (PWV) which is known to be highly correlated with blood pressure. PWV is a measure of the rate or speed at which pressure waves move down the vessel. When the heart pumps blood, it moves out of the left ventricle into the aorta where it is pushed through the rest of the circulatory system. During systole, the contraction of the left ventricle and the ejection of blood into the ascending aorta dilates the aortic wall in order to generate an aortic pressure wave which moves along the arterial tree. The velocity of this aortic pressure movement gives a measurement of arterial compliance. It has been established that PWV is a highly reliable measure for prognosis in various disease conditions such as renal disease, diabetes, and hypertension. PWV can best be measured by using two pressure catheters placed a known distance from one another. The distance between the catheters is referred to as the pulse wave distance. The PTT can be computed from the time it takes for the pressure wave to travel from the upstream pressure catheter (close to the heart) to the downstream pressure catheter (away from the heart). Then, from PTT, the PWV can be calculated by dividing the pulse wave distance by the PTT, which, in turn can be used for providing a measure of cardiovascular health. Fig. 3 shows how PTT can be measured from simultaneously-recorded PPG and ECG signals.

However, PWV can also be estimated in a noninvasive manner. Instead of placing a catheter, PTT can be measured from two different biosignals from two different peripheral body parts. ECG and fingertip PPG are most popularly used for this purpose. Please note that both signals must be synchronized in time for measuring PTT. PTT is typically measured from the time difference between the R wave on the electrocardiogram to the pulse wave arrival at the finger.

There are number of approaches that exist in available literature for measuring PTT from the simultaneously recorded PPG and ECG signal. The most popular way of measuring PTT is to measure the time difference in the R wave of the ECG signal and a point in the corresponding PPG pulse located at 50% of the pulse height. Fig. 3 shows a pictorial representation of the measurement of PTT from PPG and ECG.

Once PTT is measured, systolic and diastolic blood pressure can be directly measured. The simplest way of estimating BP is to create a curve

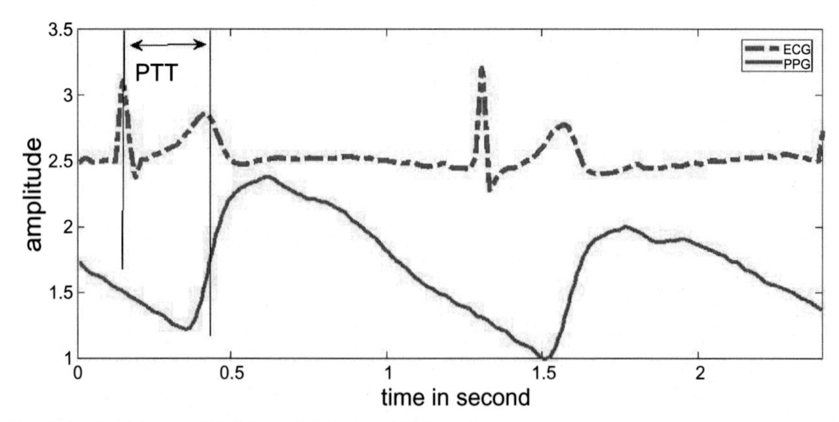

Fig. 3 Estimation of PTT from PPG and ECG.

fitting based on linear regression or nonlinear regressors based on ANNs or SVMs. In various approaches, other parameters such as heart rate, BMI, age, and PPG parameters such as average pulse width and curvature are also used as additional features along with PTT to estimate blood pressure. PTT-based approaches reported much accurate BP estimators when compared to pure PPG based approaches. Available research works have been able to estimate very accurate blood pressure (<5 mmHg error compared to ground truth value) using the regressor-based learner. In another work [21], researchers have mathematically related PTT with blood pressure to estimate systolic and diastolic blood pressure in every cardiac cycle using a pure mathematical approach. In Ref. [22], authors presented an end-to-end deep learning architecture for estimating BP from PTT.

4. Cardiac fatigue from PPG and ECG

4.1 Why investigating cardiac fatigue is important

It is well known that sedentary lifestyle and unhealthy eating habits are major causes of chronic cardiopulmonary disorders and related deaths [23]. According to the Center for Disease Control and Prevention (CDC) [24], more than 40% [25] of such disorders cause deaths outside the hospital, with failure to detect early warning signs being the main reason. Hence it becomes important to screen for abnormality in the cardiopulmonary system.

The cardiopulmonary exercise test (CPET) [26] is an important clinical tool for detection of cardiac stress levels in patients with heart and lung disease, or patients who are scheduled to undergo relevant surgeries. This test requires the patient to breathe into a specially-constructed mouthpiece with embedded sensors and the ECG of the subject is recorded before, during, and after a stationary workout such as on-the-spot cycling. Such tests are designed to provide accurate and detailed heart and lung performance of the subject while being physically stretched via active exercises. However, due to the use of specialized equipment, there are serious concerns about their practical and easy widespread deployment and use in natural day-to-day scenarios. People with no complaints or symptoms may fail to take such specialized tests at the onset of concerned disorders, which, in turn, may deteriorate with time.

Fatigue can be associated with variety of disorders. For cardiovascular diseases such as heart failure, valvular heart diseases, cardiomyopathies, and coronary artery disease, fatigue is a quite commonly observed symptom.

It is well studied in clinical body of knowledge that fatigue [27,28] often becomes the first early clinical manifestation of these diseases and as it progresses, the quality of life and prognosis deteriorates. If we investigate the pathophysiological mechanisms of fatigue in detail, we find that it is related to efficacy loss of the effector muscle. Such loss, in turn, is usually caused by mismatch of cardiac output during exercise. However, muscle and microcirculatory deconditioning, neuroendocrine dysfunction, and associated metabolic disorders also contribute toward such loss.

People, as part of their daily activity, naturally perform short spells of intensive activities in their routine, e.g., walking on stairs, brisk walking, which, in their normal course, can put their physiology through some amount of abnormal stress resulting in cardiac fatigue. According to the American College of Cardiology Foundation/American Heart Association (ACCF/AHA) [29,30], the most noteworthy and common symptoms of any cardiac disorder is shortness of breath (dyspnea) and fatigue, coupled with palpitations if the body is subjected to an above-ordinary exertion [31]. Recording major physiological parameters and studying them during these spells unobtrusively while the subject is engaged in their regular activities, can eliminate/reduce the need for specialized tests. Any anomaly or irregularity in the stress-bearing capability of the body can be detected, which in turn, can provide the signs for any early onset of cardiopulmonary disorder. The New York Heart Association (NYHA) has undertaken extensive study of this phenomenon and they try to classify subjects into different categories denoting the stages/progression of a possible cardiac disorder [32]. Here the target detection class can be taken as the subjects exhibiting dyspnea, palpitation, and fatigue in physical activities (Functional Capacity II, Objective Assessment B). These symptoms are seen to be quite prevalent among the patients developing cardiopulmonary disorders. This is reflected in the ACCF/AHA staging system also, where level C includes all the NYHA classes, and marks fatigue and dyspnea as associated symptoms [33].

4.2 Sensing cardiac fatigue

Smartwatches can be used as opportunistic ubiquitous sensing [34] of the physiological parameters toward detection of palpitation and dyspnea symptoms resulting in fatigued states during the physically intensive spells of a user. The smartwatch has an inertial measurement unit (IMU) [35] consisting of accelerometer/gyroscope and photoplethysmogram (PPG), which can be used for above measures. IMU can be used to detect the spells of intensive activities, and PPG can be used to extract cardiac and breathing cycle information. In addition to a 24×7 monitoring in a natural setting, such a system can also be used as a longitudinal clinical assessment tool. Here a practicing clinician can direct the subject to perform certain specified activities with customizable parameters such as a preworkout resting duration. Physiological parameters via these sensors can be recorded before and after the specific activity. The system can also create a questionnaire that can record additional information metadata about the subject such as current fitness level (normal, moderately sick, or acutely sick), and comfort level while doing activity. This can help in assisting the specialist doctor with in-depth analytics of the recorded parameters. Such inputs can also be used in a ML-driven model in addition to the sensor data.

The heart acts as a pump that continuously flows oxygenated blood to the vital organ. The demand of oxygen in the cells and tissues varies depending upon physical activities. When running or jogging, the internal organs need more oxygen. In this scenario, there is the need for the heart to pump more blood, thereby increasing the heart rate. However, as the heart works more beyond its capacity, it can cause tiredness, shortness of breath, and a feeling of being simply worn out. These are the signs of cardiac fatigue, which is one of the most common symptoms of congestive heart failure. Cardiac fatigue is an important parameter to estimate certain cardiac death. A certain cardiac death can happen not only to a cardiac patient but also to an apparently asymptomatic healthy person or even to an athlete. Biosignals such as PPG and ECG can be deployed for measuring and tracking various parameters that can generate useful warnings on cardiac fatigue.

The two most important parameters related to cardiac fatigue are heart rate and breathing rate. The resting heart rate of a normal person typically lies between 60 and 90 bpm. However, for an athlete, the resting heart rate can be as low as 40 bpm. During high intensity physical activity, the heart rate increases. It is normal to increase the heart rate from 60 bpm to 120 bpm during running. However, it is alarming if it reaches beyond 200 bpm. Hence, it is advisable to immediately relax if the heart rate reaches above 200 bpm. Successive R-R interval distances in ECG or the peak-to-peak interval distances in PPG can provide measurement for instantaneous heart rate. However, ECG is considered the more reliable source for accurate measurement of heart rate, and it can also track the fluctuation in instantaneous heart rate.

A person's breathing rate is also an important marker. A normal person breathes 14–18 times in 1 min under resting conditions, which significantly increases during workout. The rate of change in the breathing rate is a good indication of cardiac fatigue. A faster increase in the breathing rate is always alarming. Breathing rate can be successfully measured from the PPG. The PPG signal contains a slowly-varying low frequency drift which indicates the breathing rate. The low frequency envelope can be easily extracted by applying a low pass filter of desired frequency band. The good news is that the heart rate and the breathing rate never overlap in the same frequency region. Hence, they can be easily segregated.

Another parameter of consideration is the recovery time of the heart: when you have stopped a physical activity, how quickly does the heart rate reach the baseline resting heart rate from the elevated heart rate? A slow recovery time is a typical indication of cardiac disease, and such people are prone to cardiac arrest during a workout. The recovery time can also be easily estimated from PPG and ECG.

4.3 Some interesting early results and path to the future

In Ref. [36], the authors proposed a concept called the cardiopulmonary care platform (C2P) that can qualitatively measure cardiac fatigue using PPG. In this paper, in addition to measuring heart rate and breathing rate, the authors also calculated another parameter, called the breathing power for monitoring cardiac fatigue using a commercial-off-the-shelf (COTS) smartwatch equipped with a PPG sensor. C2P is a continuous monitoring platform for ubiquitously screening for cardiopulmonary disorders in the early-onset stage. The authors undertook an analysis of different physiological markers for fatigue from continuously monitored smartwatch data followed by a comparative study of the results toward efficacy in detecting phases of above-ordinary physical exertions. It presents a method to extract breathing cycles from PPG sensor data and devises a robust algorithm to detect spells of stairs while walking. The authors claim 85%–98% accuracy in detecting four types of activity—stationary, walking, climbing stairs, going down stairs. They also show reasonable accuracy in detecting breathing rate from fatigue. Finally, they show how heart rate and breathing rate changes before and after different types of activities followed by how a novel feature like breathing power can be used as a marker for cardiac fatigue.

Such work can be extended, by correlating the fatigue markers with the metabolic equivalent of task (MET) [37,38] of activities for each subject and allowing the platform to detect any symptoms of unhealthy fatigue. This can be achieved by creating and deploying a system for longitudinal data collection and clinical analysis. In such a system, a specialist can analyze a subject's past and current physiological data along with automated understanding of the current activities to generate fatigue markers and early warning systems for future clinical assessment.

5. Conclusion

It is well known that hypertension or elevated blood pressure is a critical risk factor of various cardiac conditions. Although the elderly population is more vulnerable to form hypertension, it may affect young adults as well. If detected, medication, a healthy lifestyle, and modified diet can help to keep the blood pressure in control, but hypertension cannot be cured. Hence, hypertension patients are regularly required to monitor their blood pressure. Although a mercury sphygmomanometer is the most accurate tool for measuring blood pressure, it cannot be used for continuous monitoring. In this chapter, we have shown different techniques for noninvasive monitoring of blood pressure indirectly from biosignals such as PPG and ECG. Such technologies are even deployed in modern smart commercial appliances for monitoring blood pressure. However, we need to keep in mind that such devices are used for approximate estimation of blood pressure and hence they cannot be considered as a direct replacement of the mercury sphygmomanometer. As these solutions involve a training phase to estimate the blood pressure, their performance widely depends upon the diversity of the training data. Moreover, the solutions are often required to be recalibrated frequently with respect to a sphygmomanometer.

In this chapter, we have also discussed how we can measure cardiac fatigue from PPG and ECG. Cardiac fatigue is an important marker toward determination of the cardiac risk factor which needs to be regularly monitored during exercise and other types of physical stress. In this chapter we outline how use of wearable/portable sensors can help toward such measurements. In particular, we discuss how the CPET, the clinical gold standard in this space, can be approximated by continuous monitoring of daily activities via the inertial measurement unit (IMU) of smartphones and cardiac parameters such as breathing power can be measured from PPG before and after activity. Breathing power is shown to a reliable marker for cardiac fatigue. Early results for such a wearable based monitoring system are also presented.

In the future, such ubiquitous monitoring of hypertension and cardiac fatigue may lead to the personalized longitudinal trend monitoring of people using wearables, heading toward creating personalized digital twins of cardiovascular systems [39,40]. Such systems can help in detection of early onset of cardiovascular abnormalities and can have a disruptive effect in cardiovascular management.

References

[1] Joint National Committee on Detection, Treatment of High Blood Pressure, and National High Blood Pressure Education Program. Coordinating Committee, Report of the joint national committee on detection, evaluation, and treatment of high blood pressure, National Heart, Lung, and Blood Institute, National High Blood Pressure Education Program, 1995.

[2] P.K. Elias, R.B. D'Agostino, M.F. Elias, P.A. Wolf, Blood pressure, hypertension, and age as risk factors for poor cognitive performance, Exp. Aging Res. 21 (4) (1995) 393–417.

[3] N. Westerhof, J.-W. Lankhaar, B.E. Westerhof, The arterial windkessel, Med. Biol. Eng. Comput. 47 (2) (2009) 131–141.

[4] A. Tsanas, J.Y. Goulermas, V. Vartela, D. Tsiapras, G. Theodorakis, A.C. Fisher, P. Sfirakis, The Windkessel model revisited: a qualitative analysis of the circulatory system, Med. Eng. Phys. 31 (5) (2009) 581–588.

[5] Y.-H. Kao, T. Tse-Yi, P.C.-P. Chao, Y.-P. Lee, C.-L. Wey, Optimizing a new cuffless blood pressure sensor via a solid–fluid-electric finite element model with consideration of varied mis-positionings, Microsyst. Technol. 22 (6) (2016) 1437–1447.

[6] T. Hanazawa, K. Asayama, D. Watabe, A. Tanabe, M. Satoh, R. Inoue, A. Hara, et al., Association between amplitude of seasonal variation in self-measured home blood pressure and cardiovascular outcomes: HOMED-BP (Hypertension Objective Treatment Based on Measurement By Electrical Devices of Blood Pressure) Study, J. Am. Heart Assoc. 7 (10) (2018) e008509.

[7] N. Bessonov, A. Sequeira, S. Simakov, Y. Vassilevskii, V. Volpert, Methods of blood flow modelling, Math. Model. Nat. Phenom. 11 (1) (2016) 1–25.

[8] M. Abdi, M. Navidbakhsh, A. Razmkon, A lumped parameter method to calculate the effect of internal carotid artery occlusion on anterior cerebral artery pressure waveform, J. Biomed. Phys. Eng. 6 (1) (2016) 33.

[9] Y. Shi, P. Lawford, R. Hose, Review of zero-D and 1-D models of blood flow in the cardiovascular system, Biomed. Eng. Online 10 (1) (2011) 1–38.

[10] S.N. Doost, D. Ghista, S. Boyang, L. Zhong, Y.S. Morsi, Heart blood flow simulation: a perspective review, Biomed. Eng. Online 15 (1) (2016) 1–28.

[11] Z. Duanmu, W. Chen, H. Gao, X. Yang, X. Luo, N.A. Hill, A one-dimensional hemodynamic model of the coronary arterial tree, Front. Physiol. (2019) 853.

[12] X. Zhang, W. Dan, F. Miao, H. Liu, Y. Li, Personalized hemodynamic modeling of the human cardiovascular syste a reduced-order computing model, IEEE Trans. Biomed. Eng. 67 (10) (2020) 2754–2764.

[13] F. Lamonaca, K. Barbe, Y. Kurylyak, D. Grimaldi, W. Van Moer, A. Furfaro, V. Spagnuolo (Eds.), Application of the artificial neural network for blood pressure evaluation with smartphones, 2013 IEEE 7th International Conference on Intelligent Data Acquisition and Advanced Computing Systems (IDAACS), vol. 1, IEEE, 2013, pp. 408–412.

[14] A. Visvanathan, R. Banerjee, A.D. Choudhury, A. Sinha, S. Kundu, Smart phone based blood pressure indicator, in: Proceedings of the 4th ACM MobiHoc Workshop on Pervasive Wireless Healthcare, 2014, pp. 19–24.

[15] O. Schlesinger, N. Vigderhouse, D. Eytan, Y. Moshe, Blood pressure estimation from PPG signals using convolutional neural networks and Siamese network, in: IEEE International Conference on Acoustics, Speech and Signal Processing (ICASSP), 2020, pp. 1135–1139.
[16] A. Tazarv, M. Levorato, A deep learning approach to predict blood pressure from ppg signals, in: 43rd Annual International Conference of the IEEE Engineering in Medicine & Biology Society (EMBC), 2021, pp. 5658–5662.
[17] A.D. Choudhury, R. Banerjee, A. Sinha, S. Kundu, Estimating blood pressure using Windkessel model on photoplethysmogram, in: 2014 36th Annual International Conference of the IEEE Engineering in Medicine and Biology Society, IEEE, 2014, August, pp. 4567–4570.
[18] S. Datta, R. Banerjee, A.D. Choudhury, A. Sinha, A. Pal, Blood pressure estimation from photoplethysmogram using latent parameters, in: 2016 IEEE International Conference on Communications (ICC), IEEE, 2016, May, pp. 1–7.
[19] R. Banerjee, A. Ghose, A.D. Choudhury, A. Sinha, A. Pal, Noise cleaning and Gaussian modeling of smart phone photoplethysmogram to improve blood pressure estimation, in: 2015 IEEE International Conference on Acoustics, Speech and Signal Processing (ICASSP), IEEE, 2015, April, pp. 967–971.
[20] R.P. Smith, J. Argod, J.-L. Pépin, P.A. Lévy, Pulse transit time: an appraisal of potential clinical applications, Thorax 54 (5) (1999) 452–457.
[21] X. Ding, B.P. Yan, Y.-T. Zhang, J. Liu, N. Zhao, H.K. Tsang, Pulse transit time based continuous cuffless blood pressure estimation: a new extension and a comprehensive evaluation, Sci. Rep. 7 (1) (2017) 1–11.
[22] H. Eom, D. Lee, S. Han, Y.S. Hariyani, Y. Lim, I. Sohn, K. Park, C. Park, End-to-end deep learning architecture for continuous blood pressure estimation using attention mechanism, Sensors 20 (8) (2020) 2338.
[23] Center for Disease Control and Prevention, National Center for Health Statistics. https://www.cdc.gov/nchs/fastats/deaths.htm. retrieved 11 February 2022.
[24] Centers for Disease Control and Prevention. https://www.cdc.gov/. retrieved 11 February 2022.
[25] Center for Disease Control and Prevention, National Center for Health Statistics, Retrieved July 24 https://www.cdc.gov/heartdisease/facts.htm. retrieved 11 February 2022.
[26] K. Albouaini, M. Egred, A. Alahmar, D.J. Wright, Cardiopulmonary exercise testing and its application, Postgrad. Med. J. 83 (985) (2007) 675–682.
[27] R. Nelesen, Y. Dar, K. Thomas, J.E. Dimsdale, The relationship between fatigue and cardiac functioning, Arch. Intern. Med. 168 (9) (2008) 943–949.
[28] G. Claessen, A. La Gerche, Exercise-induced cardiac fatigue: the need for speed, J. Physiol. 594 (11) (2016) 2781–2782.
[29] Home—American College of Cardiology. https://www.acc.org/. retrieved 11 February 2022.
[30] American Heart Association, To Be a Relentless Force for a World of Longer, Healthier Lives. https://www.heart.org/. retrieved 11 February 2022.
[31] M. Jessup, W.T. Abraham, D.E. Casey, A.M. Feldman, G.S. Francis, T.G. Ganiats, M. A. Konstam, D.M. Mancini, P.S. Rahko, M.A. Silver, 2009 focused update: ACCF/AHA guidelines for the diagnosis and management of heart failure in adults, Circulation 119 (14) (2009) 1977–2016.
[32] American Heart Association, Classes of Heart Failure. http://www.heart.org/HEARTORG/Conditions/HeartFailure/AboutHeartFailure/Classes-of-Heart-Failure_UCM_306328_Article.jsp. retrieved 11 February 2022.
[33] Heart Failure Criteria, Classification, and Staging. http://emedicine.medscape.com/article/2500037-overview. retrieved 11 February 2022.

[34] A. Khasnobish, R. Rakshit, A. Sinharay, T. Chakravarty, Demo Abstract: Phase-gain IC based novel design of tidal breathing pattern, sensor for pulmonary disease diagnostics, in: 16th ACM/IEEE International Symposium on Information Processing in Sensor Networks (IPSN), April 2017.
[35] What Is an Inertial Measurement Unit? VectorNav. https://www.vectornav.com/resources/inertial-navigation-articles/what-is-an-inertial-measurement-unit-imu. retrieved 11 February 2022.
[36] V. Chandel, S. Mukhopadhyay, D. Jaiswal, D.S. Jani, S. Khandelwal, A. Pal, C2p: an unobtrusive smartwatch-based platform for automatic background monitoring of fatigue, in: Proceedings of the First International Workshop on Human-Centered Sensing, Networking, and Systems, 2017, pp. 19–24.
[37] Using Metabolic Equivalent for Task (MET) for Exercises. https://www.verywellfit.com/met-the-standard-metabolic-equivalent-3120356. retrieved 11 February 2022.
[38] M. Jetté, K. Sidney, G. Blümchen, Metabolic equivalents (METS) in exercise testing, exercise prescription, and evaluation of functional capacity, Clin. Cardiol. 13 (8) (1990) 555–565.
[39] C. Genevieve, A. Figtree Gemma, F. Fletcher David, R. Julie, The health digital twin: advancing precision cardiovascular medicine, Nat. Rev. Cardiol. 18 (12) (2021) 803–804.
[40] O. Mazumder, D. Roy, S. Bhattacharya, A. Sinha, A. Pal, Synthetic PPG generation from haemodynamic model with baroreflex autoregulation: a digital twin of cardiovascular system, IEEE EMBC (2019) 5024–5029.

CHAPTER 7

Correlated diseases

1. Introduction

Heart is one of the most vital organs of human body. It pumps oxygenated blood to the organs and the tissues. Abnormality in cardiac operation may fail to provide sufficient oxygen supply to the rest of the body, leading to malfunction of other organs and manifestation of various noncardiac diseases. Conversely, if any of the major organs such as kidney, liver, or lungs malfunction, that often puts extra pressure on the heart causes cardiac abnormalities. Let us take the example of kidney disease. A person having cardiac problem may develop CKD (chronic kidney disease) because of the reduced blood flow to the kidneys. On the other hand, for a person having CKD, kidneys are not performing well. As a result, the heart's workload increases as it needs to circulate more blood, leading to high blood pressure and heart diseases. That is why some physiological signals capable of classifying cardiac diseases also tend to offer distinguishing features for noncardiac diseases. In this chapter, we will discuss two such noncardiac diseases, namely sleep disorders and COPD (chronic obstructive pulmonary disease) which are highly correlated with cardiac diseases. Both are regarded as strong risk factors of cardiovascular events.

1.1 Sleep disorders

Sleep is an integrated part of personal health and well-being. Although it is not scientifically clear why we sleep, several studies demonstrate that sleep plays an important role in our general well-being. There are plenty of research studies indicating that poor sleep adversely affects health markers such as motor function, irritability [1], obesity [2], and depression [3,4]. Unsurprisingly, cardiovascular disease (CVD) is also strongly associated with poor sleep. Sleep apnea is an independent risk factor for hypertension, cardiac arrhythmias, and stroke. Treatment of sleep problems using continuous positive airway pressure (CPAP) reduces blood pressure, arrhythmias, and cardiac incidences leading to reduced mortality and increasing quality of life [5]. Obstructive sleep apnea (OSA) increases the risk of heart failure, stroke,

New Frontiers of Cardiovascular Screening using Unobtrusive Sensors, AI, and IoT
https://doi.org/10.1016/B978-0-12-824499-9.00007-6

and coronary heart diseases by 140%, 60%, and 30% respectively [6]. Some researchers believe that this association between OSA and CVD is not only correlative, but also causative [7]. While the underlying causation is not clearly understood, several explanations have been proposed such as prolonged sympathetic activation [8,9], modification in intrathoracic pressure [10], and oxidative stress leading to vascular inflammation from nocturnal hypoxia and reoxygenation cycles [11,12].

Sleep deprivation is noted as a public health epidemic in multiple studies and trials. Our sedentary lifestyle with increased usage of smartphones and electronic gadgets, especially in the past decade, is making the situation worse. The prevalence rate is alarming; existing in more than 25% of both the male and female population. The American Sleep Medicine (AASM) and National Sleep Foundation (NSF) recommend that adults should regularly sleep for 7–9 h per night [13]. However, a report by the US Centers for Disease Control and Prevention (CDC) reveals that more than 80 million adults in the United States have less than 7 h of sleep [14].

1.2 Chronic obstructive pulmonary disease

Another such correlated disease is chronic obstructive pulmonary disease (COPD) which indicates obstructed airflow from the lungs. Long-term exposure to gaseous substances or particles floating in the air are the main contributors of COPD. COPD is a risk factor for lung cancer independent of smoking. In 2020, COPD was the third-leading cause of death in the top 10 causes of death globally, in the statistics published by the World Health Organization (WHO). COPD accounts for 3 million deaths globally [15]. The prevalence is generally higher in the age group of over 50 years. Vascular and heart diseases are the leading comorbidities observed in COPD. COPD patients often exhibit signs of pulmonary artery remodeling, leading to pulmonary hypertension. Smoking is the common risk factor of coronary heart disease and COPD. Coexistence of these two diseases worsens the prognosis of both [16].

Heart failure and COPD both cause shortness of breath. Many research studies have shown that people with COPD are more prone to heart disease. COPD is a progressive disease, and it is mostly irreversible. However, if detected early enough and managed properly, COPD patients can maintain a good quality of life.

2. Sleep analysis

There are different types of sleep studies. In this section, we focus on two leading factors of sleep disturbances: Sleep apnea and sleep arousal. Sleep apnea exists when the airways are blocked more than normal and repeatedly during sleep. This causes breathing to stop; often leading to loud snoring and daytime tiredness after a full night's sleep. Another type of sleep problem is unwanted sleep arousals. These are brief period of wakefulness while sleeping [17]. Note that some spontaneous arousals are normal features of the brain. However, unwanted excessive arousals can take place when sleep is disturbed.

2.1 Sleep studies

A polysomnography (PSG), also commonly known as a sleep study, is the current gold standard to diagnose sleep disorders. A PSG requires the subject to sleep with multiple sensors (the number of sensors ranging from 13 to 25) attached to the body. Continuous time-series data for all those sensors are recorded for the entire sleep duration. The sensors include brain activity using electroencephalography (EEG), eye movements using electrooculography (EOG), muscle movements using electromyography (EMG), oxygen saturation (SpO_2), and cardiac events (ECG). After a PSG is completed, experienced scorers (sleep technicians) review the entire data in 30-s windows commonly known as epochs. The sleep score consists of multiple information including, but not limited to, overall sleep efficiency, sleep stages, sleep arousals, breathing irregularities such as apneas and hypoapneas, limb movement, and oxygen saturation during sleep. All of these are important markers of sleep quality. For example, periodic limb movements of sleep (PLMS) are a common sleep disorder defined as repetitive, stereotypical, and unconscious leg movements during sleep. Research also reveals that PLMS is a very high-risk factor for increased CV morbidity and mortality [18]. EMG signals can precisely record these leg movements. Similarly, other sensors can record irregularity in brain waves and eye movements, changes in heart rate and breathing rate, and changes in blood oxygen levels. All these synchronous time-series data is recorded and reviewed by a polysomnography technologist, who uses the data to check for certain patterns as per the AASM Scoring manual [19]. A detailed analysis of AASM guidance is beyond the scope of this book. In the subsequent sections we will show how to automate the pattern detection in the PSG data. The technologist charts the sleep stages and sleep arousals, along with potential problems, in the final report. This complete sleep report, along with the medical history of the subject, is reviewed by a physician. Often, based on the findings, if the subject exhibits sleep apnea syndromes, the physician may suggest the person to undertake a positive airway pressure (PAP) titration study where a mask is fitted during sleep. In a PAP machine, a stream of air is fed into the nostrils using a tight-sealed nosepiece. This is done to verify whether the added external air pressure by the PAP machine is improving the disturbed parameters. For example, a severely sleep deprived person can have very low SpO_2 during sleep, say, in the range of 70–80. Let us also assume that such a person's SpO_2 is otherwise normal (say, more than 98 when awake). With proper PAP setting, the lowest SpO_2 during sleep may see a dramatic improvement (say, $SpO_2 > 90$) which indirectly confirms the cessation (or reduction) of apnea events. This ensures enough oxygen flow into lungs during sleep, resulting into a normal sleep and reduced daytime tiredness. Furthermore, this greatly reduces the long-term negative effects such as cardiovascular complications.

2.1.1 Sleep stage classification

Sleep stage classification defines the fundamental character of a sleep study. AASM recognizes five stages as follows: Wake (W), Rapid Eye Movement (REM), Non REM1 (N1), Non REM2 (N2), and Non REM3 (N3). N3 is often referred as slow wave sleep or deep sleep. As shown in Fig. 1, during a normal overnight sleeping session, the sleep stage oscillates in multiple cycles. It moves gradually from Awake to REM to N1 to N2 to N3, and then backtracks to REM. REM is the lightest form of sleep; dreams are most vivid in this stage. During REM stage, the eyes and eyelids flutter and breathing becomes irregular. On the other hand, in the non-REM sleep stages (i.e., N1, N2, and N3) muscles tend to be more relaxed, blood

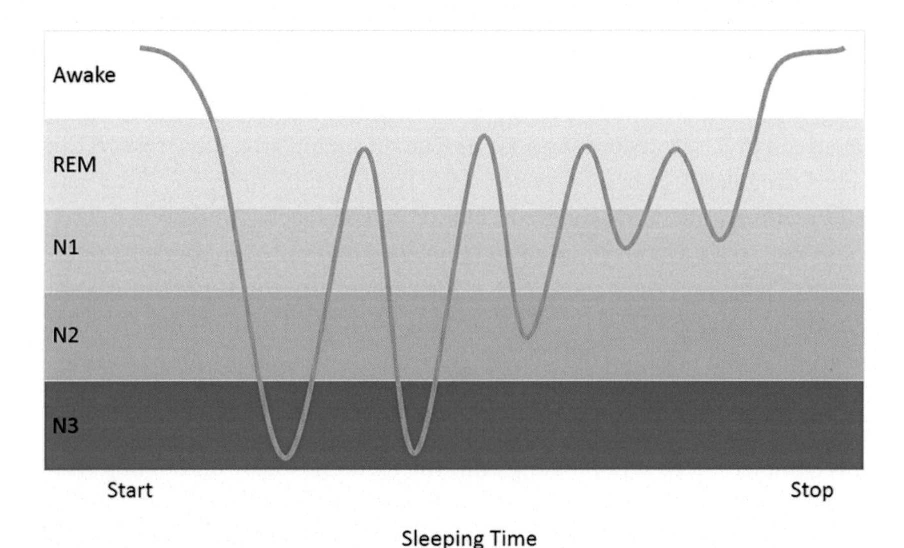

Fig. 1 A normal sleep and associated sleep stages.

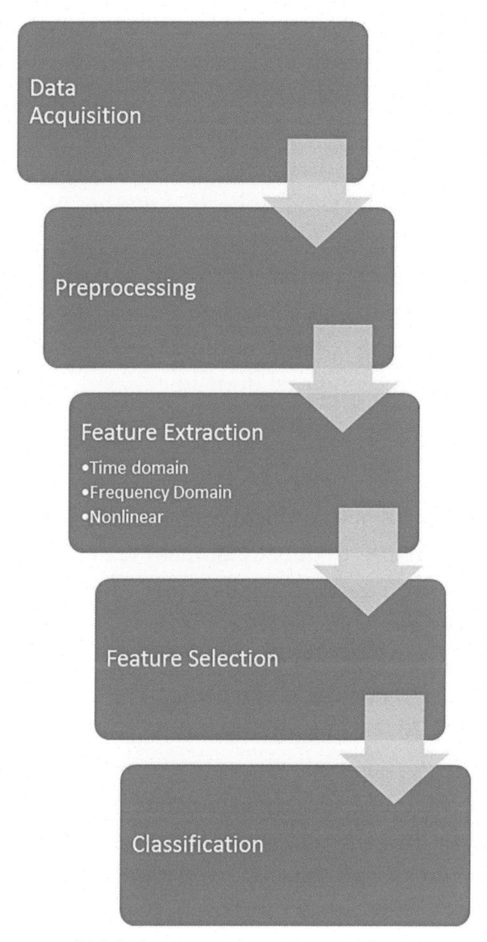

Fig. 2 Processing steps of PSG data.

pressure, breathing rate and heart rates drop, and delta wave brain activity (the slow brain activity) increases. Fig. 2 demonstrates the conventional processing steps for sleep stage classification using nondeep learning methods. In the subsequent sections, we will discuss each of these steps.

Preprocessing

Out of all the signals recorded during sleep, EEG is particularly prone to noise. Several methods are proposed in order to eliminate the effects of artifacts in EEG signals. The methods include:

- Digital filter: Different types of digital filters are use on EEG. A notch filter is often used to eliminate the power noise. An EEG consists of different frequency bands such as delta (1–4 Hz), theta (4–8 Hz), alpha

(8–12 Hz), beta (15–30 Hz), and gamma-band oscillations (>30 Hz). A bandpass (typically Butterworth) filter is used to separate these. Band pass filters are also effective to reduce Butterworth filters and linear trends.
- Wavelet transform (WT): WT decomposes a signal into a subband of frequencies. We can calculate the energy in a subband, and subsequently use them as features. Symmlet and Daubechies mother wavelets are the most common used wavelets for EEG processing.
- Independent component analysis (ICA): By thresholding the energy of the source signals, the independent signal with low energy level can be eliminated. Thereafter, a reverse transform can output the denoised signal back into the spatiotemporal domain.

Feature extraction

State-of-the-art studies suggest multiple features, which can be broadly classified into time domain and frequency domain features.
- Time domain features: Time domain features capture different morphological characteristic of the signal.
  - Statistical features: Some of the widely-used time domain features are mean, variance, skewness, and kurtosis. These are calculated on multiple signals such as EEG, EOG, and EMG.
  - Hjorth parameters: Hjorth parameters include activity, mobility, and complexity [20]. The activity parameters represent the variance of a time function, i.e., the signal power. A mobility parameter captures the proportion of the standard deviation to the power spectrum. Complexity, on the other hand, indicates the change in frequency. These parameters are very popular choice of features for EEG analysis.
  - Zero crossing: Number of times the signal has crossed the baseline. The baseline can be created using the mean value of a windowed signal.
- Frequency domain features: These features are mainly extracted from EEG signals as those contained important embedded in different frequency bands:
  - Spectral estimation: Auto-regressive (AR), moving average (MA), and auto-regressive moving average (ARMA) are the prominent parametric methods used to estimate power spectra [21]. On the other hand, periodogram and Welch methods are the popular non-parametric methods to estimate the power spectral density.

  - Higher order spectra (HOS): As the name suggests, the HOS represents the higher order information of the signals. HOS goes beyond the power spectrum which is second-order information. A third-order spectrum, termed bi-spectrum, is proposed in Ref. [22].
- Time-frequency domain features: EEG is nonstationary signal and hence it yields interesting features in the time-frequency domain.
  - Signal decomposition: Short time Fourier transformation (STFT) is a simple signal decomposition method, where signal is often divided into overlapping windows and then frequency transformation is applied to each such window. Similarly, wavelet transform is another popular method that applies different filters generated from the chosen mother wavelet.
  - Energy distribution: Hilbert Huang transform (HHT) uses empirical mode decomposition (EMD) to decompose the signal into intrinsic mode function (IMF). Then Hilbert spectral analysis (HSA) is applied to identify the instantaneous frequency.
- Nonlinear features: Entropy-based features capture the irregularity and impurity of the signal. Different flavors of entropy have been proposed in the literature such as Renyi's entropy, sample entropy, Tsallis entropy, permutation entropy, Lempel-Ziv, multiscale entropy (MSE), and approximate entropy (ApEn). Some researchers proposed fractal-based features. These methods look for self-similarity by analyzing a signal in the phase space. The correlation dimension, Lyapanov exponent, and Hurst exponent are some examples used in sleep stage classification.

Feature selection and classification

Like all classification problems, a handful of feature selection methods are applied in sleep stage detection as well. The main objective is to remove redundant features and in turn reduce overfitting (Fig. 2). Sequential forward selection and sequential backward selection are the simplest methods. The number of features from the overnight sleep data of so many sensors often leads to very high dimensions. To tackle this, principle component analysis (PCA) and linear discriminant analysis (LDA) are brought into play for feature reduction.

Several classification methods are proposed in the literature. K nearest neighbor (kNN) typically stands out as a popular choice as the problem exhibits multimodal distribution. However, some researchers also propose support vector machine (SVM) methodology as it often provides more stability. For a given set of features, random forest (RF) seems to provide the highest performance compared to the rest of the classifiers. The main advantage of RF over other classifiers is the inherent ensemble nature where the input sample of trees is randomly repeated multiple times in order to reduce the effect of noise and outliers. In Table 1, we summarize some of the top performing state-of-the-art methods of sleep stage classification. Note that:

1. Different sensors have been used to detect sleep stages. However, EEG is the most common sensor, followed by ECG.

Table 1 State-of-the-art methods for sleep stage classification.

Brief methodology	Dataset	Result
Actigraphy features and dynamic warping of body movement are used to extract statistical features [23].	Actigraphy and respiratory efforts	95.7% accuracy
Single-channel EEG signals are decomposed using three level wavelet decomposition, followed by log energy, signal fractal dimensions, and sample entropy features. SVM was used as the classifier [24].	The benchmark Sleep-EDF Dataset	98.3% accuracy for two class classification
Novel time domain features termed statistical behavior of local extrema are extracted from each frequency band of EEG. Multiclass SVM is used as the classifier [25].	The benchmark Sleep-EDF Dataset	97.9% accuracy for two class classification to 90.6% classification for six class classification
Deep convolutional neural networks (CNNs) are used to extract time invariant features, frequency features, and a sequence. A sequence model was used to capture the long- and short-term dependencies from a single- channel EEG [26].	The benchmark Sleep-EDF Dataset	84.26% accuracy for two class classification

Table 1 State-of-the-art methods for sleep stage classification—cont'd

Brief methodology	Dataset	Result
RR time series extracted from ECG signals were used to generate recurrence qualification analysis and entropy features. Variance and dispersion entropy of different frequency bands of EEG are also considered as features. Finally, DNN architecture of autoencoders is used to classify the stages [27].	MIT-BIH polysomnographic dataset	95.71% accuracy for sleep vs wake

2. Nondeep learning techniques with different strategies of time domain and frequency features with a special emphasis on wavelet features have been proven to be successful.
3. Deep neural networks (DNN) are gradually becoming the preferred choice for sleep classification.

2.2 Sleep apnea and sleep arousal

Sleep apnea is quantified using multiple parameters as defined below:

- Oxygen desaturation index (ODI): Normal oxygen saturation can be as low as 96% to 97%. Up to 90% is mildly abnormal, below 90% is considered moderately abnormal, and if SpO_2 falls below 80% the situation is highly abnormal. Nocturnal hypoxemia is often characterized by a low oxygen saturation level. The ODI is defined as the count of episodes in 1 h of sleep where the SpO_2 fell below 3% of the corresponding baseline of the concerned subject.
- Apnea-hypoapnea index (AHI): AHA is the count of apnea and hypoapnea events per hour in one session of sleep. The ratings of AHA is marked as normal ($AHA < 5$), mild ($5 \leq AHI \leq 15$), moderate ($15 \leq AHI \leq 30$), and severe ($AHI \geq 30$).
- Respiratory disturbance index (RDI): RDI represents the respiratory difficulties per hour of sleep in one session of sleep. In addition to apneic and nonapneic events, respiratory effort related arousals (RERAs) cause

respiratory disturbances. RDI is expressed as number of (apneas + hypoapneas + RERAs)/sleep hours.

Sleep arousals are brief episodes of wakefulness. The PhysioNet Challenge 2018 dataset has a collection of 13 signals with annotations of RERA arousals and non-RERA, nonapnea arousals [28]. The objective of the challenge was to detect arousals on unseen test data after the training had been undertaken on the training data. The data was partitioned to ensure that the train and test data groups had a similar distribution of AHIs. The drug usage patterns of the subjects who underwent the sleep test were also noted. It is interesting that the hypertension drugs ranked highest (41%) among the entire drug history of the subjects [28], which highlights a strong association of cardiac problems with sleep arousals.

2.2.1 Detection techniques

Typical approaches to detect sleep apnea and sleep arousal in a ML-based workflow include the following steps: Dividing the entire signal in small overlapping windows, calculating the time domain and frequency domain features, followed by training using popular ML techniques such as SVM, k-NN, ensemble learning, decision trees, AdaBoost, and LogitBoost. Some of the methods used parameter tuning [29–31].

Given the huge size of the Physionet Challenge dataset (13 sensors, each with 7–9 h of recording at 200 Hz), classical ML method proved to be extremely time-consuming. Sleep analysis is an ideal candidate for time-series deep neural network (DNN). Recurrent neural networks (RNN) exploits the persistence of past information by adding loops in the learning. They achieve this by connecting the nodes to form a directed graph along a temporal sequence. Simple vanilla RNNs often suffer from a vanishing gradient problems. This issue occurs when the network fails to propagate meaningful gradient information from the output end of the network back to the input end. As a result, learning RNNs on a long data sequence such as sleep data becomes hampered.

Long short-term memory (LSTM) is a special type of RNN which takes care of the vanishing gradient problem. It employs three gates: Input, output, and forget. The forget gate helps LSTM to forget information upon availability of new information. Researchers have successfully detected obstructive sleep apnea (OSA) using LSTM on instantaneous heart rate (IHR) and SpO_2 signals [32]. IHR can be calculated using ECG signals. The cardiac interval is a quasiperiodic signal. The irregularity of heart rate, i.e., heart rate variability (HRV) is captured in the IHR signal.

3. COPD

Chronic obstructive pulmonary disease (COPD) is a very common disease affecting 2%–3% of the global population [33]. The primary causes of COPD are smoking and air pollutants. The symptoms include shortness of breath, tiredness, and coughing. The death rate is also very high, especially in lower- income countries [34]. COPD is often reported along with a few other comorbidity conditions. Hypertension, diabetes, chronic heart failure, and coronary heart diseases are the most frequently reported among such conditions [35]. Hence, it becomes extremely important to detect COPD at an early phase, in order to find the related cardiac ailments. Moreover, COPD is an irreversible condition causing difficulty in breathing. It can only be managed with certain lifestyle changes as well as medicine, thereby enhancing both life expectancy and quality of life.

The computer science fraternity has been investigating different modalities of COPD detection for the past few decades. While most of the techniques are aimed at classifying COPD using automated analysis of lung sounds, other novel approaches exist. Texture quantification of computed tomography (CT) images is a popular problem in image processing forums [36]. Researchers also applied ML techniques on permittivity biosensors used to characterize the dielectric properties of human saliva [37]. Exhaled volatile organic compounds (VOCs) are another form of sensors successfully deployed in identifying COPD patients [38]. Although VOC samples can be collected using a portable device, nonportable methods such as gas chromatography and mass spectroscopy are required to extract actionable information from the data. Hence, although we see a handful of novel directions explored in cutting edge research, not all of them are user-friendly or easy to deploy in a home scenario. In this book, we restrict our analysis to those sensors that can be realized in a typical home environment. Hence, we would limit our discussion of this chapter to the analysis of time-series signals such as lung sounds and different metadata such as age, weight, height, and gender.

Spirometry is the gold standard in the diagnosis of COPD. Spirometry measures the lung volume and airflow speeds during inhalation and exhalation. Different early symptoms may be developed during COPD or other obstructive lung diseases (OLDs). However, spirometry, although very effective, is an elaborate process and often requires multiple attempts from the subject to complete successfully. On the other hand, general physicians rely on a common set of symptoms such as lasting cough, breathing difficulties, crackles, wheezing, and nasal allergy. A prominent differential diagnosis of COPD is asthma. However, it often requires differential diagnosis by a specialist physician to distinguish between them [39]. Given this background, it is important to have a low-cost automatic lung sound screening tool.

Auscultation, a process invented by Rene Laennec in 1816, is the process of listening to heart sounds. Two centuries down the line, although auscultation remains the primary mode of pulmonary diagnosis, it is still a manual process. The results are largely subject to the clinician's hearing capability and subjective interpretation of sounds. Wheezes and crackles are the most commonly encountered sounds found in auscultation of COPD patients. A wheeze is a continuous abnormal sound occurring in a narrow airway. In a frequency-time plot as detailed in Fig. 3, a wheeze demonstrates continuous high energy throughout the low frequencies (<1 kHz) typically lasting for a few seconds. On the other hand, crackles are a discontinuous rattling sound. The crackles may cover a wider frequency range of frequencies. In the frequency-time plot, it often shows as short bursts (i.e., spikes) of energy.

The most common research efforts collect lung sounds from the human chest using a digital microphone. In Ref. [40], the authors proposed wavelet packet transform (WPT), followed by classification using artificial neural network (ANN).

Automated lung sound analysis has been a well-known research topic for decades [41]. However, in this book we limit our discussion to COPD-related efforts [42–44]. Unfortunately, there is no uniform standardized protocol for collecting data. Some data uses state-of-the-art collected sound data from one fixed body position, whereas a few have more than one microphone. In the next sections, we will discuss using conventional ML as well as deep learning to classify COPD using lung sound analysis.

3.1 Conventional machine learning

The flow of applying conventional machine learning methods for COPD detection is no different from the steps shown in Fig. 2. We elaborate each of the processing as steps as follows [43]:

1. Preprocessing: Nyquist's theorem states that a periodic signal must be sampled at least twice at the highest frequency component of the signal. As all the adventitious lung sounds lie within 0-–2 kHz [41], the lung sound can be safely downsampled to 4 kHz as per the Nyquist theorem. Because the heart and lungs both make distinct sounds, it is a common

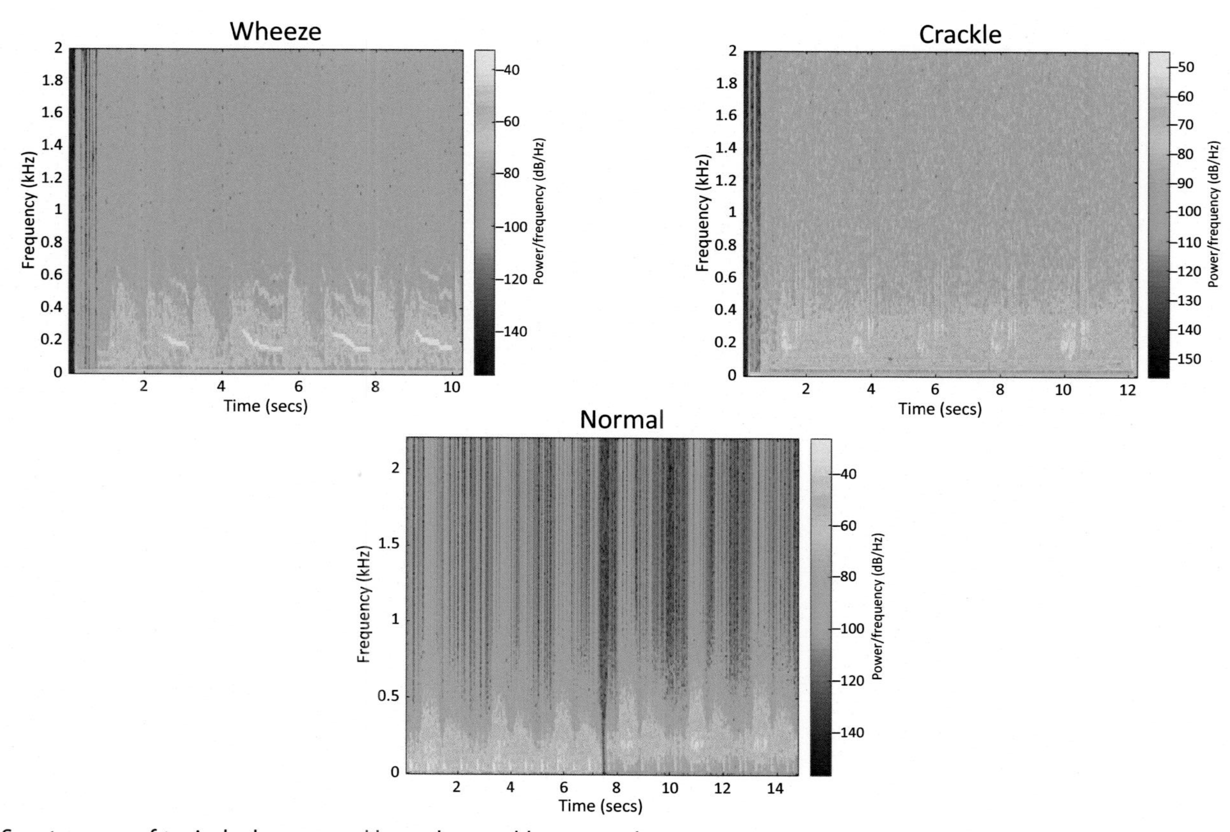

Fig. 3 Spectrogram of typical wheeze, crackle, and normal lung sounds.

phenomenon to capture both during recording. However, as we want to analyze as much pure lung sound as possible, we need to remove the heart sound from the recorded data. Empirical mode decomposition is one of the most popular methods to achieve this.

2. Feature extraction: Various features can be extracted from the processed signals. The same can be further categorized as follows:
   - **(a)** Spectral features: The Welch method can be used to generate the power spectral density (PSD) of the overlapping segments. Normal breathing sounds generally have a sharp peak at a low frequency (say, 50–150 Hz) which trails off greatly before 500 Hz. On the other hand, an abnormal lung sound will typically be spread over a broader frequency band. Different features of the PSD plot can be used to differentiate these variations in PSD plots. For example, ratios of low (<500 Hz) and high frequency (>500 Hz) power, centroid, flux, roll-off and kurtosis are potential features.
   - **(b)** Spectrogram features: Normalized power content of small frequency windows can be used as features. These give us more control to track the power over smaller frequency bands.
   - **(c)** Wavelet features: Discrete wavelet transform (DWT) provides the advantage of representing both temporal and frequency domain information. Approximation and detail coefficients of the audio signal in the region of interesting frequencies, i.e., 0–2000 Hz can be calculated and used as features [45].
   - **(d)** Cepstral features: Mean and standard deviation of the Mel frequency cepstral coefficients (MFCC) and linear frequency cepstral coefficients (LFCC) can be used as features. These prove to be discriminative in differentiating crackle and wheezes [46].
3. Feature selection: It is important to filter out the redundant features, hence maximal information coefficient (MIC)-based features with ranking algorithm can be used to choose the top features.
4. Once features are selected, common neural networks such as SVM or random forest can be used to classify the audio files.

3.2 Deep learning

In Ref. [42], the authors presented a semisupervised method to detect wheezes and crackles. They used a Hamming window filter in the order of 200 with a cutoff value of 1600 Hz. The audio was downsampled to 4 kHz. Each audio file was converted to a spectrogram using short time

Fourier transform (STFT). Interestingly, they used the ML technique generally employed for image classification on this two-dimensional STFT output. A denoising autoencoder (DA) with three layers of 50 neurons was used to train on a mix of labeled and unlabeled data. Then the top 10 features of the DA were selected for wheezes and crackles respectively. Finally, SVM was used for the classification task. The accuracies of crackle and wheeze detection were 74% and 86% respectively.

In Ref. [47], authors proposed using deep belief networks (DBN) with the Hilbert-Huang statistical features to classify COPD. They claimed detection accuracy of 70% which jumped to 90% once a sequential forward feature selection was applied to the DBN.

In Ref. [48], authors selected breath cycles and computed a 256 point PSD for each cycle. Then they applied a genetic algorithm to select 129 data values and applied them to neural network. Finally, wheezes and crackles were classified using multilayer perceptron running backpropagation algorithm. The classification performance was 93.8%.

In Ref. [49], authors used transfer learning for lung sounds classification. The dataset had four different types of sounds: Normal, wheeze, crackle, and rhonchi. However, their aim was to differentiate normal from abnormal, effectively making it a two-class classification. Lung sounds were converted into two-dimensional image-like representations using Mel spectrograms. They tried multiple per trained image feature extractors such as VGG16, IceptionV3, DenseNet201, ResNet50, and ResNet101 with two classifiers, CNN and SVM. The features extracted by VGG16 yielded the best performance with both the classifiers. However, CNN consistently performed better than SVM for all the combinations. The reported accuracy is 85%, which the authors claimed to be better than the average medical professionals.

4. Conclusion

In this chapter, we demonstrated that the time-series signals typically monitored in cardiovascular domain can also be analyzed to detect other chronic diseases and health markers. We covered sleep disorders and COPD as examples and described the processing steps for each of them. The idea presented in this chapter is predominantly beneficial for two reasons: Lowering the cost of 24×7 surveillance of a patient by using low-cost sensors, and the use of sensors already present in our daily gadgets such as smartphones and smartwatches, thereby leading to a possibility of large-scale ubiquitous deployment. Given the plethora of sensors becoming mainstream by appearing in handheld devices, we believe that this approach will be extended to other diseases in the future.

References

[1] J.J. Pilcher, A.I. Huffcutt, Effects of sleep deprivation on performance: a meta-analysis, Sleep 19 (4) (1996) 318–326.

[2] R.P. Ogilvie, S.R. Patel, The epidemiology of sleep and obesity, Sleep Health 3 (5) (2017) 383–388.

[3] D. Nutt, S. Wilson, L. Paterson, Sleep disorders as core symptoms of depression, Dialogues Clin. Neurosci. 10 (3) (2008) 329.

[4] M. Lee, A.C. Choh, E.W. Demerath, K.L. Knutson, D.L. Duren, R.J. Sherwood, S.S. Sun, et al., Sleep disturbance in relation to health-related quality of life in adults: the Fels Longitudinal Study, JNHA J. Nutr. Health Aging 13 (6) (2009) 576–583.

[5] G. Jean-Louis, F. Zizi, L.T. Clark, C.D. Brown, S.I. McFarlane, Obstructive sleep apnea and cardiovascular disease: role of the metabolic syndrome and its components, J. Clin. Sleep Med. 4 (3) (2008) 261–272.

[6] E. Shahar, C.W. Whitney, S. Redline, E.T. Lee, A.B. Newman, F. Javier Nieto, G.T. O'Connor, L.L. Boland, J.E. Schwartz, J.M. Samet, Sleep-disordered breathing and cardiovascular disease: cross-sectional results of the Sleep Heart Health Study, Am. J. Respir. Crit. Care Med. 163 (1) (2001) 19–25.

[7] Y. Peker, L. Grote, H. Kraiczi, J. Hedner, Sleep apnea a risk factor of cardiovascular disease, Lakartidningen 99 (45) (2002) 4473–4479.

[8] E.C. Fletcher, Cardiovascular disease associated with obstructive sleep apnea, Monaldi Arch. Chest Dis. 59 (3) (2003) 254–261.

[9] E.C. Fletcher, Sympathetic over activity in the etiology of hypertension of obstructive sleep apnea, Sleep J. Sleep Sleep Disord. Res. (2003).

[10] J.D. Parker, D. Brooks, L.F. Kozar, C.L. Render-Teixeira, R.L. Horner, T.D. Bradley, E.A. Phillipson, Acute and chronic effects of airway obstruction on canine left ventricular performance, Am. J. Respir. Crit. Care Med. 160 (6) (1999) 1888–1896.

[11] L. Lavie, R. Lotan, I. Hochberg, P. Herer, P. Lavie, A.P. Levy, Haptoglobin polymorphism is a risk factor for cardiovascular disease in patients with obstructive sleep apnea syndrome, Sleep 26 (5) (2003) 592–595.

[12] L. Lavie, Oxidative stress in obstructive sleep apnea and intermittent hypoxia–revisited–the bad ugly and good: implications to the heart and brain, Sleep Med. Rev. 20 (2015) 27–45.

[13] V.K. Chattu, M.D. Manzar, S. Kumary, D. Burman, D.W. Spence, S.R. Pandi-Perumal, The global problem of insufficient sleep and its serious public health implications, in: Healthcare, vol. 7, no. 1, Multidisciplinary Digital Publishing Institute, 2019, p. 1.

[14] Consensus Conference Panel: N.F. Watson, M.S. Badr, G. Belenky, D.L. Bliwise, O. M. Buxton, D. Buysse, et al., Joint consensus statement of the American Academy of Sleep Medicine and Sleep Research Society on the recommended amount of sleep for a healthy adult: methodology and discussion, J. Clin. Sleep Med. 11 (8) (2015) 931–952.

[15] The Top 10 Causes of DEATH. Global Health Estimates by World Health Organisation. https://www.who.int/news-room/fact-sheets/detail/the-top-10-causes-of-death. (Date last accessed 19 February 2022).

[16] A. Cavaillès, G. Brinchault-Rabin, A. Dixmier, F. Goupil, C. Gut-Gobert, S. - Marchand-Adam, J.-C. Meurice, et al., Comorbidities of COPD, Eur. Respir. Rev. 22 (130) (2013) 454–475.

[17] M. Bonnet, D. Carley, M. Carskadon, C. Paul Easton, R. Guilleminault, B. Harper, M. Hayes, P.K. Hirshkowitz, S. Keenan, ASDA report. EEG arousals: scoring rules and examples, Sleep 15 (2) (1992) 173–184.
[18] M. Mirza, Frequent periodic leg movement during sleep is associated with left ventricular hypertrophy and adverse cardiovascular outcomes, J. Am. Soc. Echocardiogr. (2013) 783–790.
[19] R.B. Berry, The AASM manual for the scoring of sleep and associated events, Rules, Terminology and Technical Specifications, 176, American Academy of Sleep Medicine, Darien, IL, 2012, p. 2012.
[20] B. Hjorth, EEG analysis based on time domain properties, Electroencephalogr. Clin. Neurophysiol. 29 (3) (1970) 306–310.
[21] T. Kayikcioglu, M. Maleki, K. Eroglu, Fast and accurate PLS-based classification of EEG sleep using single channel data, Expert Syst. Appl. 42 (21) (2015) 7825–7830.
[22] U.R. Acharya, E.C.P. Chua, K.C. Chua, L.C. Min, T. Tamura, Analysis and automatic identification of sleep stages using higher order spectra, Int. J. Neural Syst. 20 (06) (2010) 509–521.
[23] X. Long, P. Fonseca, J. Foussier, R. Haakma, R.M. Aarts, Sleep and wake classification with actigraphy and respiratory effort using dynamic warping, IEEE J. Biomed. Health Inform. 18 (4) (2013) 1272–1284.
[24] M. Sharma, G. Deepanshu, P.V. Achuth, U. Rajendra Acharya, An accurate sleep stages classification system using a new class of optimally time-frequency localized three-band wavelet filter bank, Comput. Biol. Med. 98 (2018) 58–75.
[25] S. Seifpour, H. Niknazar, M. Mikaeili, A.M. Nasrabadi, A new automatic sleep staging system based on statistical behavior of local Extrema using single channel EEG signal, Expert Syst. Appl. 104 (2018) 277–293.
[26] S. Mousavi, F. Afghah, U. Rajendra Acharya, SleepEEGNet: automated sleep stage scoring with sequence to sequence deep learning approach, PLoS One 14 (5) (2019) e0216456.
[27] R.K. Tripathy, U. Rajendra Acharya, Use of features from RR-time series and EEG signals for automated classification of sleep stages in deep neural network framework, Biocybern. Biomed. Eng. 38 (4) (2018) 890–902.
[28] M.M. Ghassemi, B.E. Moody, L.-w.H. Lehman, C. Song, Q. Li, H. Sun, R.G. Mark, M.B. Westover, G.D. Clifford, You snooze, you win: the PhysioNet/computing in cardiology challenge 2018, in: 2018 Computing in Cardiology Conference (CinC), 2018.
[29] D.C. Mack, M. Alwan, B. Turner, P.M. Suratt, R.A. Felder, A passive and portable system for monitoring heart rate and detecting sleep apnea and arousals: preliminary validation, in: Proceedings of the 1st Transdisciplinary Conference on Distributed Diagnosis and Home Healthcare, Arlington, VA, USA, 2–4 April, 2006, pp. 51–54.
[30] R.J. Thomas, J.E. Mietus, C.-K. Peng, G. Gilmartin, R.W. Daly, A.L. Goldberger, D.J. Gottlieb, Differentiating obstructive from central and complex sleep apnea using an automated electrocardiogram-based method, Sleep 30 (2007) 1756–1769.
[31] C.-C. Hsu, P.-T. Shih, An intelligent sleep apnea detection system, in: Proceedings of the 2010 International Conference on Machine Learning and Cybernetics, Qingdao, China, 1–14 July, vol. 6, 2010, pp. 3230–3233.
[32] R.K. Pathinarupothi, J. Dhara Prathap, E.S. Rangan, E.A. Gopalakrishnan, R. Vinaykumar, K.P. Soman, Single sensor techniques for sleep apnea diagnosis using deep learning, in: 2017 IEEE International Conference on Healthcare Informatics (ICHI), 2017, pp. 524–529, https://doi.org/10.1109/ICHI.2017.37.
[33] GBD 2015 Disease and Injury Incidence and Prevalence Collaborators, Global, regional, and national incidence, prevalence, and years lived with disability for 310 diseases and injuries, 1990–2015: a systematic analysis for the Global Burden of Disease

Study 2015, Lancet 388 (10053) (2016) 1545–1602, https://doi.org/10.1016/S0140-6736(16)31678-6. PMC 5055577. PMID 27733282.
[34] Chronic Obstructive Pulmonary Disease (COPD). www.who.int. Retrieved 1 July 2021.
[35] E. Crisafulli, S. Costi, F. Luppi, G. Cirelli, C. Cilione, O. Coletti, L.M. Fabbri, E.M. Clini, Role of comorbidities in a cohort of patients with COPD undergoing pulmonary rehabilitation, Thorax 63 (6) (2008) 487–492.
[36] M. Alemzadeh, C. Boylan, M.V. Kamath, Review of texture quantification of CT images for classification of lung diseases. Critical Reviews™, Biomed. Eng. 43 (2–3) (2015).
[37] P.S. Zarrin, N. Roeckendorf, C. Wenger, In-vitro classification of saliva samples of COPD patients and healthy controls using machine learning tools, IEEE Access 8 (2020) 168053–168060.
[38] C.O. Phillips, Y. Syed, N. Mac Parthaláin, R. Zwiggelaar, T.C. Claypole, K.E. Lewis, Machine learning methods on exhaled volatile organic compounds for distinguishing COPD patients from healthy controls, J. Breath Res. 6 (3) (2012) 036003.
[39] J. Buffels, J. Degryse, J. Heyrman, M. Decramer, Office spirometry significantly improves early detection of COPD in general practice: the DIDASCO Study, Chest 125 (4) (2004) 1394–1399.
[40] A. Mondal, P.S. Bhattacharya, G. Saha, Reduction of heart sound interference from lung sound signals using empirical mode decomposition technique, J. Med. Eng. Technol. 35 (6–7) (2011) 344–353.
[41] R.X. Pramono, S.B. Adhi, E. Rodriguez-Villegas, Automatic adventitious respiratory sound analysis: a systematic review, PLoS One 12 (5) (2017) e0177926.
[42] D. Chamberlain, R. Kodgule, D. Ganelin, V. Miglani, R.R. Fletcher, Application of semi-supervised deep learning to lung sound analysis, in: 2016 38th Annual International Conference of the IEEE Engineering in Medicine and Biology Society (EMBC), IEEE, 2016, August, pp. 804–807.
[43] S. Datta, A.D. Choudhury, P. Deshpande, S. Bhattacharya, A. Pal, Automated lung sound analysis for detecting pulmonary abnormalities, in: 2017 39th Annual International Conference of the IEEE Engineering in Medicine and Biology Society (EMBC), IEEE, 2017, July, pp. 4594–4598.
[44] C. Jácome, A. Marques, Computerized respiratory sounds in patients with COPD: a systematic review, COPD J. Chron. Obstruct. Pulmon. Dis. 12 (1) (2015) 104–112.
[45] M.A. Tocchetto, A.S. Bazanella, L. Guimaraes, J.L. Fragoso, A.J.I.P.V. Parraga, An embedded classifier of lung sounds based on the wavelet packet transform and ANN, IFAC Proc. 47 (3) (2014) 2975–2980.
[46] N. Sengupta, M. Sahidullah, G. Saha, Lung sound classification using cepstral-based statistical features, Comput. Biol. Med. 75 (2016) 118–129.
[47] G. Altan, Y. Kutlu, N. Allahverdi, Deep learning on computerized analysis of chronic obstructive pulmonary disease, IEEE J. Biomed. Health Inform. 24 (5) (2019) 1344–1350.
[48] İ. Güler, H. Polat, U. Ergün, Combining neural network and genetic algorithm for prediction of lung sounds, J. Med. Syst. 29 (3) (2005) 217–231.
[49] Y. Kim, Y.K. Hyon, S.S. Jung, S. Lee, G. Yoo, C. Chung, T. Ha, Respiratory sound classification for crackles, wheezes, and rhonchi in the clinical field using deep learning, Sci. Rep. 11 (1) (2021) 1–11.

Further reading

American Sleep Disorders Association, Arousals: scoring rules and examples: a pre-liminary report from the sleep disorders atlas task force of the American Sleep Disorders Association, Sleep 15 (2) (1992) 173–184.

R. Boostani, F. Karimzadeh, M. Nami, A comparative review on sleep stage classification methods in patients and healthy individuals Comput, Methods Prog. Biomed. 140 (2017) 77–91.

G.M. Pellegrino, M. Corbo, F. Di Marco, Effects of air stacking on dyspnea and lung function in neuromuscular diseases, Arch. Phys. Med. Rehabil. (2021), https://doi.org/10.1016/j.apmr.2021.01.092. S0003-9993(21)00185-4.

S. Gyulay, L.G. Olson, M.J. Hensley, M.T. King, K.M. Allen, N.A. Saunders, A comparison of clinical assessment and home oximetry in the diagnosis of obstructive sleep apnea, Am. Rev. Respir. Dis. 147 (1993) 50–53.

A. Dunai, I. Mucsi, J. Juhász, M. Novák, Obstructive sleep apnea and cardiovascular disease, Orv. Hetil. 147 (48) (2006) 2303–2311.

SECTION 3

Future challenges

CHAPTER 8

Looking at the future

1. Introduction

In the preceding chapters, we learnt the importance of physiological sensing in the early detection and monitoring for cardiovascular diseases and how artificial intelligence (AI)-based analytics can help. We focused on the most prevalent cardiovascular disease conditions such as arterial blocks and abnormal heart rhythms, and generic symptomatic heart conditions such as hypertension and sleep disorders. We discussed how basic sensing using a photoplethysmogram (PPG), electrocardiogram (ECG), or phonocardiogram (PCG), followed by suitable signal processing and machine learning (ML) can help us identify early onset of these conditions. This in turn can help the physician and patient to work out necessary lifestyle changes for preventive healthcare.

There have been two pioneering books by Dr. Eric Topol, himself a practicing cardiologist: *The Patient will see you now: The Future of Medicine Is in Your Hands* [1] and *Deep Medicine: How Artificial Intelligence Can Make Healthcare Human Again* [2], which introduces this type of AI-based analytics of physiological sensor data to screen early onset of chronic condition from a clinician's perspective. In his first book in 2016 [1], he talks about the use of smartphones and other ubiquitous sensing devices to create a more pervasive yet affordable healthcare for all.

However, as we understand the true capabilities of AI in real world deployments, it is understood that standalone AI cannot deliver clinical grade service due various issues in accuracy, liability, bias, and explainability. He elucidates this very well in his book published in 2019 [2], where he explains why the way of the future is the practice of an AI-aided human in loop medicine.

In this chapter, we first outline the trends in noninvasive physiological sensing, where we cover flexible electronics-based wearables, implantables, ingestibles, and injectibles, body-centric communication, radar based in-body imaging, and nanobiosensing. Then we discuss the trends in AI and analytics in the form of AutoML, EdgeAI, explainable AI, and the privacy,

New Frontiers of Cardiovascular Screening using Unobtrusive Sensors, AI, and IoT
https://doi.org/10.1016/B978-0-12-824499-9.00008-8

trust, and transparency challenges thereof. Finally, inspired by the vision of the *Deep Medicine* book [2], we try to give a peek into the future as to how cardiovascular healthcare will look like in 2030—both from the patient's and the doctor's perspective.

2. Trends for physiological sensing

2.1 Advances in noninvasive physiological sensing

2.1.1 Flexible electronics-based wearables

The rigid form factor of traditional electronics precludes some of the special and innovative usage of wearables. A new area of flexible electronics is coming up to address this gap through novel materials and manufacturing processes. The market for healthcare products containing flexible electronics, as forecasted by IDTechEx [3,4], will be worth over US$8.3 billion by the year 2030. The use of flexible electronics in foldable smartphones is already becoming prevalent.

Flexible and stretchable displays and sensors allow wearables to be creative in design and form factor, yet thin, lightweight, and robust against motion/fitment artifacts, especially for epidermal wearables used for physical, physiological, and chemical monitoring. It works well for physical sensing such as tactile/movement sensors and physiological sensors such as electrocardiogram (ECGs) and temperature. Examples of such systems are wearable tattoos/patches/smartclothing for cardiac health monitoring which may include monitoring of human activity, ECG, and sweat sensing [5,6]. Flexible electronics comprises of a substrate, an active layer, and interface layers. The inorganic nanomaterials-based active layers provide good sensitivity and structural strength and usually are suited for physical sensing. On the other hand, liquid metals can provide particularly good conductivity, flexibility, and stretchability, and are suitable for physiological sensing. There is also a need for chemical sensing at a molecular level from the body to enable sweat-sensing and analysis. Such systems are usually based on microfluidics sensing, which are devices that exploit the physical and chemical properties of liquids and gases at a microscale using a lesser volume of samples, chemicals, and reagents [7]. These have application in monitoring secondary cardiac health parameters in the form of dehydration monitoring, drug monitoring, and blood glucose monitoring.

The main advantage of such flexible electronics-based sensing is enabling 24×7 high quality data collection that is free of noise artifacts that arise from movement and noncontact. Applications can range from cardiac monitoring, screening, diagnostics, and treatment compliance. However, such devices need to be designed to be safe and comfortable for long-term wear. As outlined earlier, there are two types of usage of flexible electronics: e-Tattoo or skin patches, and e-textile or smartclothing.

Electronic skin patches are wearable devices that are attached to the skin via adhesive and have the flexibility/stretchability to follow the surface of the skin. *E*-textiles take advantage of the fact that our body is in contact with textiles we wear almost all the time. They may not provide the immunity from the noise artifacts that patches provide, but they are very convenient and comfortable to wear and may have easier regulatory challenges.

2.1.2 Implantables, ingestibles, and injectibles

As wearables move toward flexible form factors, the newer in-body medical devices on the horizon are implantables, ingestibles, and injectables. All of these sense inside the body and send sensed information back using wireless communication. Implantables are usually surgically implanted while the mode of deployment for ingestibles and injectables are somewhat obvious from their names.

An example of an implantable in cardiac health is an implantable loop recorder (ILR) [8] that is similar to a pacemaker, but directly senses the electrical impulses from the heart tissue. They play a key role in early detection of atrial fibrillation (AF) in high-risk patients. Ingestible devices [9] are camera-enabled capsule devices that are swallowed like regular pills; usually they are useful in gastrointestinal diagnostics, replacing traditional methods such as endoscopy and colonoscopy. They can also monitor reactions to drugs which can be important for some of the cardiac conditions. Injectable devices [9] are microdevices that are injected into the bloodstream and may often have robotic properties. They are used for continuous monitoring of blood parameters and for neurostimulation. One probable future application in cardiac health for injectables can be in the form of clot/arterial block removers and stent deployment.

Although there are privacy concerns for such invasive devices, they may play a key role in medical prognosis and treatment. In addition to creating such systems that are safe for the body, the main challenges for such devices are still limited battery life and a challenging wireless communication environment under low power constraints [10].

2.1.3 Using the human body as a communication medium

In order to address the challenges of wireless communication mentioned in a previous chapter, some of the new research is suggesting sending signals directly using the human body as a conducting medium [11]. This can also alleviate privacy concerns and harmful effects of in-body wireless radiation.

But the human body, unlike metal conductors, is a poor conductor of electrical signals, although the conductivity improves through electrolytes in the water in the interstitial fluid in our body. An electrical impulse sent by an in-body device can cause small electrical currents in different parts of the body, which can be picked up by placing sensitive electrodes on the skin on a device such as a smartwatch. Such impulses can carry vital information.

Because the interstitial fluid is present in the bloodstream in the whole body, it can enable communication between to in-body or on-body devices quite effectively. But the main challenges of such systems are their extremely low signal strength which results in poor signal-to-noise ratios (SNRs) at the receiver. The received signal strength also fluctuates with the physiological changes in the body, some of which can occur quite frequently.

2.1.4 Photoacoustic and hyperspectral sensing

Photoacoustics is the method for measuring the acoustic responses of cells and tissues that are excited by a light source, such as a laser, that can contain both morphological and functional information. Absorption of light in a tissue causes local heating, leading to a thermoelastic expansion–contraction cycle, which in turn can generate acoustic signals. If extremely narrow pulses of light are used for excitation, usually ultrasound signals are generated that can propagate via the body tissues. The time of flight (ToF) and amplitude of ultrasound signals provide information about the location and nature of the tissue. Photoacoustics is being considered as a powerful tool to study eye problems including retinal properties and for malignant tissue detection. Their application in cardiovascular health is still under exploration [12], holding great promise in understanding blood flow and chemistry in a noninvasive manner that can include detection of sugar and other molecules in the blood. Using an array of sensors and applying computational imaging techniques, one can also build two-dimensional or three-dimensional images of the body's internal organization in a noninvasive way, although their application is limited by the ability of light to penetrate the body.

Hyperspectral imaging is another emerging imaging modality not only for disease diagnosis but also for image-guided surgery. It provides a spatial image response for a large number of frequencies. Such a response is usually modulated by material properties such as composition and texture and can provide information about tissue composition and morphology; this involves scattering and absorption in hemoglobin, melanin, and water. Reflected, fluorescent, and transmitted light can be used. Raman spectroscopy is a widely used technique which is becoming prevalent for such analysis. The measurement of the shift in wavelength of the scattered radiation can provide information regarding the chemistry and structure [13]. Raman signatures are weak in strength, however, the strength can be amplified significantly by enhancing the substrate with target-specific nanoparticles.

Hyperspectral imaging modalities can be of many types depending upon the acquisition mode, spectral range/resolution and spatial resolution, measurement mode, dispersion characteristics, and sensitivity to detector arrays. Image analysis methods need preprocessing, feature extraction and selection, and classification methods. The spectrum covered can range from ultraviolet (UV) to visible to near-infrared (NIR) regions.

Although hyperspectral imaging is mostly used for tissue imaging and surgical guidance, it has application in cardiovascular diseases. It has been shown that hyperspectral imaging can be used for detection of in vivo atherosclerotic plaques [14–16], leading toward early diagnosis of ischemic heart disease. It can be also used for surgical ablation of heart cells for treating severe AF [17].

2.1.5 Radar sensing and computational imaging

The penetrative properties of microwave frequencies enable human in-body imaging by means of a microwave ultra-wideband (UWB) radar [18]. Radar imaging can be used for simple fitness and cardiac health monitoring using mobile devices. It can also be used for subcutaneous tissue reconstruction which can help in imaging the cardiovascular circulatory system and heart. This can be thought of as an in-home alternative to computed tomography (CT) scans. Subcutaneous fat tissue measurement can provide real-time monitoring for body weight management, which is so important for cardiac health. Such imaging techniques used on the heart can also help in easy, unobtrusive, noncontact measurement of vital signs such as heartbeats and breathing, which in turn can help in cardiac disease prognosis. Radar-based imaging has the potential to be affordable, and easy to deploy at home and at the point of care.

Existing radar techniques such as synthetic aperture radar (SAR) or optical imaging techniques such as CT can be used to create images from signals reflected from internal organs. Specialized techniques such as radar waveform design, active beam forming using small antennas, and compressive sensing techniques followed by computational imaging techniques can be used to reconstruct the in-body image to the desired resolution [19].

2.2 Nanobiosensing

Nanotechnology, coupled with biotechnology, is driving the design and development of a new class of sensors called nanobiosensors. Nanobiosensors are usually built using a biological recognition molecule embedded into the surface of a signal transducer. Nanobiosensors are applied widely for molecular detection of biomarkers associated with a disease [20]. The use of a high surface area yet lightweight nanomaterials such as carbon nanotubes has helped in building more sensitive yet responsive systems.

A biosensor is a device that can detect or measure the presence of a biochemical molecule. The basis modality of such sensing is affinity-based. They use specialized probes that selectively capture and bind the molecule under consideration, called the target or analyte. The selective binding changes some physical property of the surface which now needs to be sensed via different methods such as biosensors requiring light (e.g., surface plasmon resonance [SPR] or fluorescence), mechanical motion (e.g., quartz crystal microbalance or resonant cantilever), or magnetic particles.

One of the most frequently used techniques for interrogating sensors is electrical. Electrical biosensors measure currents and/or voltages to detect the binding. They usually cost less, consume less power, and are a smaller size. Hence, they can be used for point-of-care disease diagnostics.

Biorecognition molecules, such as an antibody or DNA, are used in affinity biosensors. They selectively bind to an analyte. A signal transducer is used to sense the binding reaction. They usually have specificity and high affinity of target antibodies. This allows the biosensors to analyze complex samples at low concentrations. Given the wide possibility of applications, nanobiosensors will definitely help in cardiovascular disease diagnosis and management. Some of the most relevant and promising approaches include sensing of breath and sensing of sweat for suitable biomarkers such as troponin, whose abundance in the blood indicates an injury to the heart muscle cells.

Nanoscale affinity biosensors design has two major investigation areas: To reduce the detection limit, and increase the ability to detect several analytes in the same sample, also known as multiplexing. Reliable diagnosis of disease requires identification of the molecular marker levels, which can be achieved by such multiplexing. It should be noted here that detection of multiple analytes in one sample with varying concentration is a well-known data science problem called disaggregation [21]. AI/ML based techniques can help in such scenarios.

2.3 Genomic analytics

Doctors and physicians are now convinced that cardiovascular diseases are heritable [22,23]. There are two factors that determine the cardiac risk factor of a person. One of them is of course related to the lifestyle of the person. A sedentary lifestyle, lack of physical exercise, and an unhealthy diet are considered the primary risk factors for developing cardiac issues at an early age. Other factors such as heavy smoking and mental stress are also considered direct factors for cardiac risk. The abovementioned factors are specific to a patient's lifestyle and can be modified. However, heredity and family history also determine an important role in this matter. It is widely believed by cardiologists that someone with a past family history of chronic cardiac disease, including myocardial infarction (MI), congenital heart disease, or hypertension, are at a higher risk of developing such conditions, even after maintaining a healthy lifestyle.

Over the past few decades, scientists have been working to understand the genes and specific DNA sequence variants in human body responsible for this heritability, with the main objective being correlating genotype with phenotype. The objective is to identify the genes/DNA sequence variants that are responsible for human trait variation. Naturally-occurring genetic variations have the potential to reveal biological mechanisms of disease causation in humans. This study is particularly important for cardiovascular diseases such as MI, and chronic conditions such as hypertension, cardiac arrhythmias, congenital heart disease, cardiomyopathy, high blood cholesterol, and coronary artery disease which can result in heart attack, stroke, and heart failure. It is shown that these conditions can run in families, indicating inherited genetic risk factors. Genetics can directly influence the risk for developing heart disease. It is believed that genes control every aspect of the cardiovascular system. It should be noted that a mutation even in a single gene can increase the probability of developing heart disease. It can change the way a particular protein works in the human body and how the body processes cholesterol, thereby becoming a causal source for arteries becoming blocked. That is why even after a so-called healthy lifestyle, a person can form arterial blockages at a relatively young age through the genetic variations passed from parents to children in the DNA which is copied into every cell of the child's body [24,25]. For the same reason, other family members are encouraged to undergo screening for risk factors and early-stage asymptomatic disease, when a family member is diagnosed with heart disease or a heart disorder. Scientists have discovered six genetic variants that are believed to be directly linked to heart disease. Those variants are in or near the MTHFD1L, PSRC1, MIA3, SMAD3, CDKN2A/CDKN2B, and CXCL12 genes [22].

Genomic medicine can be a potential futuristic solution that may provide patients and their healthcare providers with the necessary genetic information that will aid in detecting coronary artery disease or other types of heart diseases at their onset, which can help in prevention and treatment.

Genetic testing may be appropriate for symptoms such as an enlarged heart, irregular heartbeat, early heart attack, coronary artery disease or stroke, untreated very high blood cholesterol, unexplained fainting with exercise or emotional stress, unexplained seizure, or seizures with a normal neurological evaluation, enlarged aorta or aortic aneurysm in the chest at young age, or sudden infant death syndrome in the family [26].

Genome editing [27] is a futuristic idea to introduce corrective mutations in the gene sequences that are protective against coronary artery disease or other type of heart disease with a one-time injection. However, most of these ideas are simply theory; substantial additional work is needed to confirm the efficacy and safety of such methods and much work is happening in this area. Addressing the underlying ethics regarding genome editing also requires complying with government regulations which are always evolving.

3. Trends for analytics and AI

3.1 Technology trends

3.1.1 AutoML

Automated ML (AutoML) is the process of automating the signal/image/vision processing pipeline. It can be thought of as automation of the manual tasks required for building a processing pipeline [28–30]. Automation can exist in feature engineering, model selection, and model hyperparameter tuning. Vendors such as Python, Amazon, Microsoft, Google, and IBM have made available a plethora of AutoML tools, e.g., Auto-sklearn, Neural Network Intelligence, AutoGluon, AutoAI, and CloudAutoML.

However, while most of these will work on text, speech, image/vision, and categorical data, they do not work well on diverse physiological sensor signals such as those needed in cardiovascular disease detection. For such sensor signals [31–33], we need a combination of both signal processing-based feature generation, with an autoencoder-based feature representation, along with automated feature selection methods. This type of AutoML system can be used to discover cardiac disease markers in physiological sensor signals. Deep learning-based systems, by their inherent design, can also automate whole-processing pipeline building [34].

3.1.2 Edge AI

There are some reasons why connected health applications need edge computing support for its AI. First, all applications that need real-time monitoring and control/low latency response cannot be interrupted by Internet delays. One such example is detection of AF from an implantable loop recorder which is then used to immediately trigger an implanted defibrillator to prevent cardiac arrest.

Second, when there is need for fusion of multisensor data that is time-sensitive or there is a cost associated to large amount of sensor data transfer to the cloud, again the fluctuating latency of Internet may cause a problem. One example of time-sensitive multisensor fusion is detection of blood pressure from ECG and photoplethysmogram (PPG) where the time delay between the peaks for ECG and PPG (also known as pulse transit time or PTT) is used to model the blood pressure.

Finally, individuals may be reluctant to transfer data to the cloud due to privacy regulations or security reasons. This is true for almost all connected health applications.

In 2021, Rubino et al. showed that extreme edge computing, where data processing happens right next to the sensor, is the most power efficient paradigm in such situations [35]. Recently, deep neural networks have demonstrated exceptionally precise results in a wide range of pattern recognition tasks, such as computer vision, natural language processing, and speech processing using dedicated hardware artificial neural network (ANN) accelerators, graphics processing units (GPUs), tensor processing units, and custom systems on chips, many of which, such as GPUs, support extensive parallelism. However, edge devices are always constrained in computing power, memory, and battery life and usually have reduced size/weight/cost, which makes deployment of AI-based analytics on edge devices a challenge [36].

In the context of such constrained low powered devices, TinyML [37] is an upcoming field of study which essentially aims to shrink these ML models in such a way that they would feasibly run on low powered devices, making them smart devices even with their constraints. The applications of this technology are endless, especially in the healthcare field. Smart healthcare monitoring devices generate huge amounts of valuable data which can be used to predict and detect ailments at an early stage. The low power and computation requirements of TinyML enable a variety of always-on use cases [38].

Let us consider a pulse oximeter device. Photoplethysmography is an optical technique that uses a low cost PPG-based monitoring device to obtain a plethysmogram. Plethysmography measures the changes in the volume of blood and is used to measure the SpO_2 levels and the heart rate of the user using the pulse oximeter device. Biomarkers such as heart rate variability (HRV) can be measured through longitudinal monitoring of this data. The main challenge is the constraints of the low-powered device to process this data. Using the longitudinal PPG data, a model can be trained using TensorFlowLite micro (TFLM) [39]. The generated model is an efficient and compressed model, known as a compact model. Using various optimized techniques, e.g., for quantization, the model is optimized for the selected embedded chipset. After that, the model is combined with the application code and compiled to generate a C++ file and a binary code. The binary code can be loaded onto the low powered device. This process can tune the microcontroller in the PPG device to use TinyML to detect HRV, which is one of the biomarkers for detecting medical conditions such as cardiovascular issues, stress and more, and can alert the user in case of exigencies. Using TinyML and embedding such compact models into low-powered devices can be the future of medicine.

In addition to using standard frameworks such as TFLM, which can provide a limited amount of model size reduction via compaction and quantization, model compression and partitioning are some of the other advanced ways to embed ML in tiny devices. However, manual compression and partitioning of ML/DL models is time-consuming and requires specialized skills and efforts. Hence the automation of such systems is of great importance. One way that model compression can be automated is by deploying novel student-teacher learning techniques and knowledge distillation where the input-output of a pretrained large-sized DNN teacher model can be approximated by a smaller-sized student model with some trade-off in model accuracy [40]. The system in Ref. [41] is a particularly good example of model compression in ECG analytics where a 10 MB model is reduced in size by more than 160 times with little degradation in accuracy using a piecewise linear model as the student network. For more complex DNNs, other options can be deployment of automated deep neural network pruning systems such as lottery ticket hypothesis (LTH) from MIT [42], which claims to have 10–100 times of model size reduction.

3.1.3 Neuromorphic computing

There are some reasons why connected health applications need edge computing support for its AI. One of the main drivers is low power computing. Neuromorphic computing will play a big role in low power, intelligent wearable, implantable, and injectable devices. For such devices, on-chip intelligence will usually negatively affect the battery life. Embedding intelligence into low-powered edge devices is seen as a solution to this challenge as it has been already shown that power efficiency increases with the proximity of the computation to the source of data. However, classical Von-Neumann architecture [43]-based systems that we use for microprocessors, microcontrollers, and GPUs are still not as efficient as mammalian brains in terms of power efficiency; the primary reason is that all these systems are clocked in nature and the memory and processing are not colocated. These traditional computing systems are thus not ideal for extreme edge scenarios because the standard von Neumann architectures inherently give rise to a memory bottleneck due to the physical separation between the processing elements and memory units.

Spiking neural networks (SNN), popularly known as the third generation of neural networks, when coupled with evolving neuromorphic hardware, seem to solve the problems associated with the clock and colocation of processing and memory [44]. SNNs are modeled as networks of more bio-plausible neurons that operate only via transmission of "spikes" or events, i.e., the processing happens asynchronously only when there is an event to process, giving rise to an inherent sparsity in processing. At the same time, their target platforms, i.e., the neuromorphic hardware are architectured as connections of neurons and synapses (in case of mammalian brains) thereby colocating processing and memory. Both factors contribute to the extreme power efficiency of SNNs running on neuromorphic hardware, estimated to be about a thousand order of magnitude less than conventional systems. For wearable devices that are constrained in terms of processing and battery life, this paradigm of neuromorphic computing using SNNs seem to offer a feasible solution due to their capability of learning from sparse data and power efficiency in processing. Going forward, there will be a host of neuromorphic processor chips that are becoming available [45].

3.1.4 Explainable AI

As described in the previous sections, AI/ML-based diagnostic and screening for cardiovascular systems look promising, but as with any AI/ML system, the output of such probabilistic systems is not 100% accurate. If they are not fully accurate, the normal human (patient in this context) may not be able to trust such an autonomous system. Hence it is imperative that output of such systems is given to an expert (doctor in this context), who in turn can interpret it and then relay the information to the patient.

However, the effectiveness of these systems is limited by a machine's current inability to explain its decisions and actions to expert human users such as doctors. This is where explainable AI (XAI) comes in; it allows domain experts such as doctors to understand, trust, and own AI-based outcomes [46]. XAI systems will have the ability to explain the rationale of their inference, highlight their shortcomings, and predict their behavior under new inputs. XAI coupled with human-computer interface (HCI) techniques can enable understandable and yet meaningful interaction dialog between a doctor and the AI system. However, sometimes explainability comes at the cost of accuracy performance and system design level trade-off decisions are needed for effective deployment [47].

Let us look at an example of XAI in the context of ECG analytics. As outlined in Chapter 4, AF detection from ECG is important. With the progress in deep learning systems, it is not that difficult to design a blackbox DNN that takes an ECG as input and provides AF/not-AF types of decisions as the output. If the DNN remains a blackbox and hence nonexplainable, a doctor cannot risk taking liability by using such inferencing for their diagnosis. On the other hand, it is well known to the doctors [48], that AF ECGs have "irregularly regular heartbeats" and have missing/noisy P-wave. It is also knowns from signal processing that Poincare plots [49] can catch "irregularly regular heartbeats" [50], and similar information can be provided by tachograms [51]. There are other methods that can catch missing/noisy P-waves [52,53]. Hence if we derive features such as Poincare plots and tachograms from the raw ECG before inputting it into a DNN system, the system becomes explainable to a certain extent.

3.2 Challenges

3.2.1 Privacy, transparency, and trust

The recent European Union General Data Protection Regulation (GDPR) [54] gives every individual a right to object to decisions made about them if it is made solely based on automated processing. They also have the right to obtain information about the existence and working of such an automated decision-making system. While the first aspect deals with pure privacy which is more regulatory in nature, the second one points toward the need for an explainable system.

The "right to explanation" [54] for an automated decision-making system can be linked to the right to obtain an "explanation of a system's functionality." It is obvious from the current state-of-the-art that it will be difficult for AI/ML-based systems to create such explainability as their outcome remains inconsistent in the sense that they vary from one test to another even if the attributes remain the same.

This brings in the other two aspects: Trust and transparency [55]. The lack of repeatability and less than perfect accuracy of AI-driven systems will make the natural instinct of the end-user of such a decision system (very often it will be the doctor) to not trust such machine inference—especially when the liability of the final decision lies with the end-user such as a doctor. The only way to build such trust is to make sure that either the AI system is explainable or its decision-making logic and algorithm tuning/training mechanism are transparent to the end-user. Such transparency of the training mechanism can also allow the end-user to examine or explore any bias that might creep in due to intended/unintended imbalance in the training data [56].

3.2.2 Security

The security of such sensing device-driven systems is driven by a range of factors. Some of the factors are highlighted below.

Address device-level security

Smart medical sensing devices, due to their embedded nature, can compromise security controls for performance, cost, and energy limitations, especially for edge computing scenarios. But interestingly, AI-based pattern recognition is being increasingly used for thread detection and identification [57].

Making AI robust to security hacks

Artificial intelligence systems can be attacked via the main livelihood of AI systems—the data. Adversaries can manipulate AI systems and alter their behavior by manipulating the data used by the AI system for training. The end goal can be malicious. While traditional cyberattacks are usually caused by "bugs" or human mistakes in code, AI attacks are usually triggered by limitations in the underlying AI algorithms in form of training or bias.

The ML algorithms learn by extracting patterns from data. These patterns are usually tied to higher-level concepts, which can be a label. An example can be a learning "irregularly regular" pattern of heartbeats in ECG to flag for AF. If such pattern can be synthetically introduced in the ECG by hacking into the system, the analyzing AI can be fooled to think that the person is undergoing AF. In a highly sophisticated scenario of a person having an automatic in-body defibrillator which is triggered to give an electrical shock to stop heart fibrillation, such a hack can be used to fatally harm the person.

Implementing AI in accordance with HIPAA compliance

While implementing AI-driven screening/diagnosis/monitoring systems, HIPAA [58] compliance is essential. HIPAA is a series of federal regulatory standards that outline the lawful use and disclosure of protected health information in the United States. At a high level, HIPAA mandates the following:

- Access management and governance/control of the data: This is needed for both actual patient data used for inferencing and the data that is used for training. Especially if the analytics platform resides outside the hospital premises, this becomes of paramount importance.
- Data encryption: HIPAA mandates the use of encryption technology by healthcare organizations in order to protect sensitive patient information. To protect against unauthorized access, organizations should adopt data encryption and secure access control technologies. One interesting recent development is the rise of the homomorphic encryption techniques [59] which allows data analytics in the encrypted domain itself.
- Data bias: Due to higher statistical prevalence of certain diseases among certain demographics, there is a huge risk that incorrectly trained algorithms will demonstrate behavior that is biased toward certain demographics. If such biased algorithms [60] are used to compute prevalence/risk and associated premiums thereof for medical insurance, the pitfalls of such systems are obvious.
- Federated learning: Here the learning algorithm itself is shared among different data sources instead of the data itself [61]. The data does not

leave the hospitals – the AI models are trained locally within each hospital. The locally trained models are federated to create a globally trained model.

4. Future vision for cardiovascular health

4.1 A day in the life of a patient in 2030

Let's imagine a typical middle-aged person, Mr. X, aged 45 years, who lives with his wife of similar age, teenage child, and father Mr. Y, who is aged 75 years in 2030. X is healthy, exercises regularly, and seemingly has no cardiac problems other than being mildly diabetic and a family history of heart ailments.

One day, X wakes up in the morning feeling uncomfortable. However, his smartmattress with pressure sensors has already figured out that he had a restless sleep and sends a message to his mirror in the bathroom. The mirror has an in-built camera and radar array that can not only analyze his facial expression and facial blood flow, but also can do in-body imaging using radar to create a CT scan similar to a 3D image for his heart and main arteries using microwave signals that are nonionizing and hence safe for use at home. It also samples his genes via his toothbrush that has an in-built genomic sensor.

A deep learning-based image analysis system analyzes the scanned image and finds some abnormality. To be sure, it instructs the smartwatch that X is wearing to take a single-lead ECG and analyzes the ECG together with the in-body scan image. The analysis is pointing toward some heart blockage and hence the AI-driven cardiac screening software advises him to see a doctor.

After getting consent from X and his wife, the software books an appointment with a cardiologist in the hospital and also orders a drone-based air ambulance to take X to hospital. X arrives in the hospital in 30 min and the cardiologist sees him along with all the measured data from home devices that have been automatically sent to the hospital over the Internet while making the appointment. The cardiologist thinks that the screening software is right, but he does some confirmatory tests such as echocardiograms and CT angiography which confirms a small block in the coronary artery.

The doctor thinks that the case has been detected early enough to be controlled by medicine and lifestyle change and he prescribes the same to X. Without the intelligent home sensing screening system via smartmattress, smartmirror and smartwatch, the heart blockage in X could have grown larger before being detected and that could have resulted in larger problems needing intervention surgery and other procedures. As X is now designated an elevated risk patient, the doctor prescribes him to also wear a tattoo-like wearable patch on his chest that constantly monitors his cardiac condition and alerts the doctor if there is something wrong.

On the hand, Mr. Y, father of X, has chronic heart rhythm problem. He had an implantable loop recorder (ILR) [62] implanted inside his chest 10 years previously. It continuously monitors his ECG via electrodes implanted inside his heart muscle. In case of any AF occurring in his ECG, an AI-driven intelligent yet battery friendly system in his ILR will pick up the AF abnormality and send it to his doctor. There is a defibrillator jacket [63] already kept in the home. The doctor will ask Y to wear it and then remotely give controlled defibrillation shocks to Y via the jacket that can control the AF and prevent Y's heart from into cardiac arrest.

The type of systems described above may sound like science fiction, but given the advances of sensing and AI, such systems are extremely plausible.

4.2 A day in the life of a cardiologist in 2030

Dr. Z is a busy cardiologist. He wakes up in the morning—his home also has similar sensors to Mr. X like a smartmattress, smartmirror, and smartwatch to monitor and screen for his health. As he gets ready to go to the hospital, his AI-driven digital personal assistant plans for his day and books an autonomous taxi.

While the doctor is traveling to the hospital, the large screen in the backseat of his autonomous taxi brings up a dashboard of his patients, their health status, and mail. For noncritical cases, he logs into the body area network of a patient, and does some finetuning of devices. He figures out that one patient might have a heart attack so he remotely readjusts the setting of defibrillator. That patient can be Mr. Y as described above.

Once he reaches hospital and enters a surgery room, he examines the patients and programs the surgical robots as per observations. There are multiple such robots in the surgery room, he monitors and controls multiple surgeries from a central workstation. In 3 h, he can perform robotic surgeries on 15 patients.

He receives an urgent call for a critical patient where a team of doctors in another hospital is doing surgery. He puts on his virtual reality gear and virtually enters the remote surgery room. He looks at patient's vitals and the current state of the surgery, and takes control of the Da Vinci surgical system [64] connected over an extremely high speed and extremely low latency 10g wireless network. He navigates to an extremely critical part of the sinoatrial node and undertakes an ablation. He then hands over control to the local team.

In the meantime, local robots have finished the operations of many people. He examines the overall progress and satisfactory completion of surgeries and asks nurses to undertake the finishing touches to surgical locations.

It is evening time, and his AI-enabled digital personal assistant knows about his energy levels and fatigue levels. It books a taxi for him to go to a spa and also books a spa appointment. Dr. Z destresses in the spa and reaches home refreshed and rejuvenated. His bed, laced with sensors senses his parameters, adjusts the temperature of the bed and the air-conditioner, and helps him to fall asleep quickly.

A day like the above is not far away for the doctors. With such tools in hand, and routine jobs taken care of by robots and AI-driven systems, doctors can focus on areas that really need their attention, such as prescribing precision-measured and personalized medicine, handling complex comorbidities, and focusing on long-term wellness outcomes by looking at disease-as-a-continuum instead of trying to fix illnesses as they usually do today.

References

[1] E. Topol, The Patient Will See You Now: The Future of Medicine Is in Your Hands, Basic Books, 2016.
[2] E. Topol, Deep Medicine: How Artificial Intelligence Can Make Healthcare Human Again, Basic Books, 2019.
[3] N. Tsao, J. Hayward, R. Das, Flexible Electronics in Healthcare 2020-2030—Printed and flexible electronics for electronic skin patches, e-textiles, test strips, and smart packaging in healthcare, 2020, IDC TechEx Report https://www.idtechex.com/fr/research-article/flexible-electronics-are-the-future-in-wearable-health-monitoring/20573. (Accessed 10 September 2021).
[4] N. Tsao, Flexible Electronics Are the Future in Wearable Health Monitoring, May 2020, IDC TechEx Research Article https://www.idtechex.com/en/research-report/flexible-electronics-in-healthcare-2020-2030/731. (Accessed 10 September 2021).
[5] W. Gao, H. Ota, D. Kiriya, K. Takei, A. Javey, Flexible electronics toward wearable sensing, Acc. Chem. Res. 52 (3) (2019) 523–533.
[6] Y. Gu, T. Zhang, H. Chen, et al., Mini review on flexible and wearable electronics for monitoring human health information, Nanoscale Res. Lett. 14 (2019) 263.
[7] W. Jing, G. Min, Microfluidic sensing: state of the art fabrication and detection techniques, J. Biomed. Opt. 16 (8) (2011) 080901.
[8] V. Vilcant, O. Kousa, O. Hai, Implantable Loop Recorder (Updated 2021 Jul 31), StatPearls Publishing, 2021. Available fro https://www.ncbi.nlm.nih.gov/books/NBK470398/.

[9] A. Kiourti, K.S. Nikita, A review of in-body biotelemetry devices: implantables, ingestibles, and injectables, IEEE Trans. Biomed. Eng. 64 (7) (2017) 1422–1430.
[10] I. Pedersen, A. Iliadis (Eds.), Embodied Computing: Wearables, Implantables, Embeddables, Ingestibles, The MIT Press, 2020.
[11] S. Sen, S. Maity, D. Das, Turning the body into a wire—when the Human Body Is the Communications Channel, it's Hard to Hack the Data, IEEE Spectr. 57 (2020) 44–49.
[12] R.D. Glickman, Photoacoustic imaging and sensing: a new way to see the eye, J. Ocul. Pharmacol. Ther. 37 (3) (2021).
[13] Raman Spectroscopy—A New Dawn in Clinical Diagnosis, Oxford Instruments, https://andor.oxinst.com/learning/view/article/raman-spectroscopy. (last accessed 10 September 2021).
[14] G. Lu, B. Fei, Medical hyperspectral imaging: a review, J. Biomed. Opt. 19 (1) (2014) 10901.
[15] J.M. Amigo (Ed.), Hyperspectral imaging in medical applications, in: Data Handling in Science and Technology, vol. 32, Elsevier, 2020, pp. 523–565 (Chapter 3.6).
[16] Medical-Biotech Applications, HeadWall Inc., https://www.headwallphotonics.com/medical-biotech. (last accessed 10 September 2021).
[17] D.A. Gil, L.M. Swift, H. Asfour, N. Muselimyan, M.A. Mercader, N.A. Sarvazyan, Autofluorescence hyperspectral imaging of radiofrequency ablation lesions in porcine cardiac tissue, J. Biophotonics 10 (8) (2017) 1008–1017.
[18] X. Zhuge, T.G. Savelyev, A.G. Yarovoy, L.P. Ligthart, J. Matuzas, B. Levitas, Human body imaging by microwave UWB radar, in: European Radar Conference, 2008, pp. 148–151.
[19] A. Kletsov, A. Chernokalov, A. Khripkov, J. Cho, S. Druchinin, Microwave noncontact imaging of subcutaneous human body tissues, Healthc. Technol. Lett. 2 (5) (2015) 108–111.
[20] V. Naresh, N. Lee, A review on biosensors and recent development of nanostructured materials-enabled biosensors, Sensors (Basel) 21 (4) (2021) 1109.
[21] S. Prasad, Nanobiosensors: the future for diagnosis of disease? Nanobiosensors Dis. Diagn. 3 (2014) 1–10.
[22] M. Kelly, C. Semsarian, Multiple mutations in genetic cardiovascular disease: a marker of disease severity? Circ. Cardiovasc. Genet. 2 (2) (2009) 182–190.
[23] A.V. Khera, S. Kathiresan, Genetics of coronary artery disease: discovery, biology and clinical translation, Nat. Rev. Genet. 18 (6) (2017) 331–344.
[24] S. Kathiresan, D. Srivastava, Genetics of human cardiovascular disease, Cell 148 (6) (2012) 1242–1257.
[25] F. Ahmad, E.M. McNally, M.J. Ackerman, L.C. Baty, S.M. Day, I.J. Kullo, P.C. Madueme, et al., Establishment of specialized clinical cardiovascular genetics programs: recognizing the need and meeting standards: a scientific statement from the American Heart Association, Circ. Genomic Precis. Med. 12 (6) (2019) e000054.
[26] https://www.uhhospitals.org/Healthy-at-UH/articles/2020/02/how-your-genes-can-influence-your-heart-health. (last accessed 7 February 2022).
[27] A. Strong, K. Musunuru, Genome editing in cardiovascular diseases, Nat. Rev. Cardiol. 14 (1) (2017) 11–20.
[28] C. Thornton, F. Hutter, H.H. Hoos, K. Leyton-Brown, Auto-WEKA: combined selection and hyperparameter optimization of classification algorithms, in: KDD '13 Proceedings of the 19th ACM SIGKDD International Conference on Knowledge Discovery and Data Mining, 2013, pp. 847–855.
[29] F. Hutter, R. Caruana, R. Bardenet, M. Bilenko, I. Guyon, B. Kegl, H. Larochelle, AutoML 2014 @ ICML, in: AutoML 2014 Workshop @ ICML, 2014.
[30] H.C. Peng, F. Long, C. Ding, Feature selection based on mutual information: criteria of max-dependency, max-relevance, and min-redundancy, IEEE Trans. Pattern Anal. Mach. Intell. 27 (8) (2005) 1226–1238.

[31] S. Banerjee, T. Chattopadhyay, A. Pal, U. Garain, Automation of feature engineering for IOT analytics, ACM SIGBED Rev. 15 (2) (2018) 24–30.
[32] A. Ukil, I. Sahu, C. Puri, A. Mukherjee, R. Singh, S. Bandyopadhyay, A. Pal, Auto-Modeling: integrated approach for automated model generation by ensemble selection of feature subset and classifier, in: International Joint Conference on Neural Networks (IJCNN), 2018.
[33] A. Ukil, P. Malhotra, S. Bandyopadhyay, T. Bose, I. Sahu, A. Mukherjee, L. Vig, A. Pal, G. Shroff, Fusing features based on signal properties and TimeNet for time series classification, in: European Symposium on Artificial Neural Networks, Computational Intelligence and Machine Learning (ESANN), 2019.
[34] A. Ukil, S. Bandyopadhyay, A. Pal, Sig-R 2 ResNet: residual network with signal processing-refined residual mapping, auto-tuned L 1-regularization with modified Adam optimizer for time series classification, in: International Joint Conference on Neural Networks (IJCNN), IEEE, 2020, pp. 1–8.
[35] A. Rubino, M. Payvand, G. Indiveri, Ultra-low power silicon neuron circuit for extreme-edge neuromorphic intelligence, in: 26th IEEE International Conference on Electronics, Circuits and Systems (ICECS), 2019, pp. 458–461.
[36] W. Jun, Edge AI Is the Next Wave of AI. Why Do You Need to Know About Edge AI, Towards Data Science, 2020. https://towardsdatascience.com/edge-ai-is-the-next-wave-of-ai-a3e98b77c2d7. (last accessed 11 February 2022).
[37] P.P. Ray, A review on tinyml: state-of-the-art and prospects, J. King Saud Univ. Comput. Inf. Sci. (2021). Elsevier.
[38] M. Shafique, T. Theocharides, V.J. Reddy, B. Murmann, TinyML: current progress, research challenges, and future roadmap, in: 58th ACM/IEEE Design Automation Conference (DAC), 2021, pp. 1303–1306.
[39] TensorFlow Lite for Microcontrollers. https://www.tensorflow.org/lite/microcontrollers. (last accessed 11 February 2022).
[40] I. Sahu, A. Pal, A. Ukil, A. Majumdar, Compressing deep neural network: a black-box system identification approach, in: International Joint Conference on Neural Networks (IJCNN), 2021.
[41] A. Ukil, I. Sahu, A. Majumdar, S.C. Racha, G. Kulkarni, A.D. Choudhury, S. Khandelwal, A. Ghose, A. Pal, Resource constrained CVD classification using single lead ECG on wearable and implantable devices, in: IEEE EMBC, 2021.
[42] J. Frankle, M. Carbin, The lottery ticket hypothesis: finding sparse, trainable neural networks, in: International Conference on Learning Representations (ICLR), 2018.
[43] Von Neumann Architecture—An Overview, ScienceDirect Topics, https://www.sciencedirect.com/topics/computer-science/von-neumann-architecture. (last accessed 11 February 2022).
[44] J.H. Lee, T. Delbruck, M. Pfeiffer, Training deep spiking neural networks using backpropagation, Front. Neurosci. 10 (2016) 508.
[45] C.D. Schuman, T.E. Potok, R.M. Patton, J.D. Birdwell, M.E. Dean, G.S. Rose, J.S. Plank, A survey of neuromorphic computing and neural networks in hardware, arXiv preprint arXiv:1705.06963, 2017.
[46] M. Turek, Explainable Artificial Intelligence (XAI), Defense Advanced Research Projects Agency (DARPA), Govt. of USA, 2018. https://www.darpa.mil/program/explainable-artificial-intelligence. (last accessed 10 September 2021).
[47] R. Schmelzer, Understanding Explainable AI, Forbes, 2019. https://www.forbes.com/sites/cognitiveworld/2019/07/23/understanding-explainable-ai/?sh=78293d7a7c9e. (last accessed 10 September 2021).
[48] Atrial Fibrillation Topic Review, Healio—Learn the Heart, https://www.healio.com/cardiology/learn-the-heart/cardiology-review/topic-reviews/atrial-fibrillation. (last accessed 10 September 2021).

[49] C.-H. Hsu, M.-Y. Tsai, G.-S. Huang, T.-C. Lin, K.-P. Chen, S.-T. Ho, L.-Y. Shyu, C.-Y. Li, Poincaré plot indexes of heart rate variability detect dynamic autonomic modulation during general anesthesia induction, Acta Anaesthesiol. Taiwanica 50 (1) (2012) 12–18.
[50] J. Park, S. Lee, M. Jeon, Atrial fibrillation detection by heart rate variability in Poincare plot, Biomed. Eng. Online 8 (2009) 38.
[51] M.J. Janssen, C.A. Swenne, J. de Bie, O. Rompelman, J.H. van Bemmel, Methods in heart rate variability analysis: which tachogram should we choose? Comput. Methods Prog. Biomed. 41 (1) (1993) 1–8.
[52] A. Gladuli, et al., Poincaré plots and tachograms reveal beat patterning in sick sinus syndrome with supraventricular tachycardia and varying AV nodal block, J. Vet. Cardiol. Off. J. Eur. Soc. Vet. Cardiol. 13 (1) (2011) 63–70.
[53] H. Pürerfellner, E. Pokushalov, S. Sarkar, J. Koehler, R. Zhou, L. Urban, G. Hindricks, P-wave evidence as a method for improving algorithm to detect atrial fibrillation in insertable cardiac monitors, Heart Rhythm. 11 (9) (2014) 1575–1583.
[54] B. Goodman, S. Flaxman, European Union regulations on algorithmic decision-making and a 'right to explanation', in: 2016 ICML Workshop on Human Interpretability in Machine Learning (WHI 2016), New York, NY, 2016.
[55] E. Thelisson, Towards trust, transparency and liability in AI/AS systems, in: Proceedings of the Twenty-Sixth International Joint Conference on Artificial Intelligence (IJCAI) Doctoral Consortium, 2017.
[56] AI in Healthcare: Data Privacy and Ethics Concerns, Lexalytics, 2021. https://www.lexalytics.com/lexablog/ai-healthcare-data-privacy-ethics-issues. (last accessed 10 September 2021).
[57] N. Sahoo, How Does Artificial Intelligence Help in Data Protection and HIPAA Compliance? 2021. https://www.cpomagazine.com/cyber-security/how-does-artificial-intelligence-help-in-data-protection-and-hipaa-compliance/. (last accessed 10 September 2021).
[58] M. Comiter, Attacking Artificial Intelligence: AI's Security Vulnerability and What Policymakers Can Do About It, Belfer Center for Science and International Affairs, Harvard Kennedy School, 2019. https://www.belfercenter.org/publication/AttackingAI. (Accessed 10 September 2021).
[59] M. Ogburn, C. Turner, P. Dahal, Homomorphic encryption, Procedia Comput. Sci. 20 (2013) 502–509.
[60] Uncovering and Removing Data Bias in Healthcare, HIMSS, https://www.himss.org/resources/uncovering-and-removing-data-bias-healthcare. (last accessed 11 February 2022).
[61] C. Brogan, New AI Technology Protects Privacy in Healthcare Settings, 2021. https://www.imperial.ac.uk/news/222093/new-ai-technology-protects-privacy-healthcare/. (last accessed 10 September 2021).
[62] Implantable loop recorder, in: Cardiac Electrophysiology: From Cell to Bedside, sixth ed., 2014.
[63] LifeVest: Treatment, Types, Recovery. https://my.clevelandclinic.org/health/treatments/17173-lifevest. (last accessed 11 February 2022).
[64] Da Vinci Surgery, Da Vinci Surgical System, Robotic Technology. https://www.davincisurgery.com/da-vinci-systems/about-da-vinci-systems. (last accessed 11 February 2022).

Index

Note: Page numbers followed by *f* indicate figures, and *t* indicate tables.

P

Q

R

S

T

U

V

W

Printed in the United States
by Baker & Taylor Publisher Services